# The Definitive Real-World Guide to Mobile Solar Power Systems

*Your Complete Guide to Achieving
Truly Sustainable Mobile Off-Grid Solar Power*

By Scott A. Rossell

Skizzle Thoughts Publishing

© 2017 - Skizzle Thoughts Publishing (a division of Rossell Enterprises)

Printed in the United States of America

First edition: September 2017

ISBN-13: 978-1977714138
ISBN-10: 1977714137

All rights reserved. No part of this publication may be reproduced, stored in a retrieval system, or transmitted in any form or by any means electronic, mechanical, photocopying, recording, or otherwise, without prior written consent of Skizzle Thoughts Publishing.

Any trademarks, brand names, model names and numbers mentioned in this text belong to their respective manufactures. Neither the publisher nor the author are in any way affiliated with or compensated by these manufacturers.

## General Disclaimer
By reading this disclaimer, you assume full responsibility for the use of the materials and information contained within the pages of this book. This book does not serve to provide medical, legal, technical, or professional advice in any form. Under no circumstances will the authors or publisher be held responsible for any loss or damages caused by your use or reliance on the information published here or contained in resources related to this book including any associated websites, software, books, eBooks, documents, periodicals, audio, video, or other materials. There are no guarantees that you will earn any money or achieve any specific results using the information in this material. Your level of success in achieving the potential results discussed in this book depends on your individual commitment, skills, knowledge, experience, and/or financial resources. The authors have made the best effort to provide accurate information on the subjects discussed in light of the availability and accuracy of data prior to publishing. However, neither the author nor publisher assume any responsibility for errors or omissions that may have occurred and reserve the right to update the material subject matter as new information is made available.

# About the Author

This is where I tell you about myself in third person so I sound really impressive. You know, "Scott has a degree is blah blah blah." Or "Scott has done this or that in these really cool places."

Hey, sure that kind of stuff is important for some subjects – and solar power SHOULD be one of them. But here's the rub; you can buy books about solar power and get everything from Billy Bob the fourth and his endless yammering about what all the supposed options are and how to hook them up to the other end of the spectrum with Chief Engineer O'Gumflapper who will throw five different acronyms at you in a single paragraph and let you guess as to what the heck he's talking about.

What do you get with me? You ready for this? I'm a houseless man living in a truck in San Diego. I've been living off solar power for nearly six years now. I read everything I could find on the subject, talked to endless salespeople, electricians, professional RV solar power system installers, and read endless blog and forum pages, then tried to spend as little money as possible and fell flat on my face several times trying to make things work. And after about four years of real-life trial and error, I finally got it right.

What you get from the other guys is a lot of theory and options. What you get from me is what actually works, because I have to live day-to-day on what I can harvest from the sun.

And I do quite well. I run my laptop, iPad, iPhone, security cameras, TV, microwave oven, instant K-Cup Keurig coffee maker, air-conditioner, swamp cooler, fans, lights, Xbox 360, Blu-ray player, ice maker, toaster oven, crock pot, rice cooker, and even a bread maker – all off power from the sun! And I never need to use a generator. You can do the same. It's not that hard. Well…once you read my book, anyway.

# Acknowledgements

To my Dad, who always asked,
**"How?"**

To my mother, who was too paranoid to say,
**"Thanks."**

To my ex-wife, who asked,
**"That's it?"** ...just one too many times.

And to all of my friends who constantly ask,
**"Why Scott, WHY?!"**

# Table of Contents

| | |
|---|---|
| **INTRODUCTION** | **1** |
| **SOLAR PANELS** | **15** |
| PANEL PLACEMENT | 21 |
| MOUNTING | 23 |
| SERIES, PARALLEL OR BOTH? | 25 |
| PANEL MAINTENANCE | 27 |
| SOLAR PANELS RECAP | 28 |
| **BATTERIES** | **29** |
| FLOODED LEAD ACID BATTERIES | 29 |
| AGM OR GEL CELL BATTERIES | 34 |
| LITHIUM BATTERIES | 35 |
| 12 VOLT OR 6 VOLT? | 37 |
| MOUNTING AND PLACEMENT | 38 |
| WIRING CONFIGURATIONS | 43 |
| BATTERY CHARGE EXPECTATIONS | 46 |
| BATTERIES RECAP | 59 |
| **SOLAR CHARGERS** | **61** |
| MOUNTING AND PLACEMENT | 69 |
| SOLAR CHARGERS RECAP | 73 |
| **POWER INVERTERS** | **74** |
| SPECIFICATIONS | 78 |
| FLOATING GROUND | 79 |
| INVERTERS RECAP | 81 |

## WIRING — 82
### Which Gauge for What? — 87
### Fuses — 92
### Wiring Recap — 97

## POWER MONITORS — 99
### Ghost Loads — 103
### Power Needs Assessment — 105
### Power Monitoring Recap — 107

## DETERMINING POWER REQUIREMENTS — 108

## BATTERY CHARGERS — 112

## GENERATORS — 117

## CAPACITORS — 121

## BATTERY RESUSCITATION TRICKS — 124

## BASIC ELECTRONICS — 129

## EXAMPLE SYSTEM DIAGRAMS — 133
### Simple/Minimal System — 133
### Day-to-Day Functional System — 135
### Robust "Road Warrior" System — 137

## SOURCES FOR MATERIALS AND EQUIPMENT — 140

# Author's Disclaimer

Each state has its own weird regulations regarding how and what you can and cannot do to or inside a vehicle in order to install a mobile solar power system. Most concerns are with properly securing the solar panels to the roof and where you put the batteries, but that's about it.

I once wrote a book on the different laws governing juvenile computer hacking back in the 1980's and I included ALL of the existing laws from each and every state. BORING! Consequently, you'll have to do your own homework on what your state will allow. I suggest contacting your local Department of Motor Vehicles or Department of Transportation. But get ready for some blank stares and awkward silences over the phone because you're going to get a lot of "Ummm...." Most of them just don't know.

In California, there are pages of laws on seat belts, but not one I could find on solar panels or batteries in regards to mounting or installation on a moving vehicle. So best of luck. Regardless, I wish I didn't have to say this right off the bat, but don't come after me if you get hurt or in trouble with the law because you did something based on what you read in this book. The information provided in this book is meant solely as a suggestion of what you might do with proper guidance, training, equipment and legal counsel. It also serves as a history of my own mistakes and lessons learned from the challenges along the way on my eventual successful path to self-sufficiency. (Sakes, try saying that ten times fast.) In short, I cannot be held responsible with what you might do with this information.

# Introduction

Right away, I wanted to show you the big picture of what you're up against in terms of equipment, costs, information and complexity involved in building a proper and reliable mobile solar power system. But try not to get too bogged down in the details. Just let the story unfold as you read it. I guarantee we'll cover every little detail later in specific chapters for each aspect of the process. In fact, some issues will be repeated over and over in different contexts to make sure they sink in.

You should also know that you have just stepped into very murky and uncharted territory. You would think that by now in the twenty-first century, the solar power industry would have a solid understanding of solar power systems, but that's just not the case. Every book I could find on the subject - EVERY ONE OF THEM - was either inaccurate, incomplete, or downright misleading. Even my go-to source, the "Dummies" series of books left me hanging.

Of course, the sales people and retail companies that sell solar products will talk a big show, but when you start to pin them down on specifics, it quickly becomes obvious that very few of them have ever actually done it themselves. They never actually had to a build a functional power generation system they had to depend on and live with on a daily basis. In fact, few have even worked with products outside of their own brand. And to compound the problem, equipment designed for home is rarely the best choice for mobile installations - and few sales people know the difference.

Sure, if you look around for specialists who know a lot about batteries or solar panels or wiring or generators or inverters, you'll likely find a few, but it's extremely rare to find any one single person who knows it all. I've been fighting with the big picture strictly in a mobile setting for over five years now and I still don't have it all figured out. And what I do know as fact is often contested - but almost always by people who only sell equipment and have never actually tried living with it on a daily basis.

I have tried endless configurations and countless types of equipment to find the best solution for my needs. I have also broken and repaired a lot of stuff along the way. Almost every mistake I have made was because of inaccurate or non-existent information so I had to wing it on my own and hope for the best. There have been a lot of "gotchas" along the way. I've also spent far more money on the wrong things and far less on some of the important ones because I didn't know the difference. I fell for sales pitches and untested theories instead of tried and true advice mainly because in the last five years I have come across only TWO other people who actually knew what they were talking about. And that was because they did exactly what I did; they actually built and scaled functional mobile solar power systems that they had to rely on for their daily power needs.

In this book, you'll learn a lot...I mean, A LOT. And a good deal of what I have to teach you is what NOT to do; what not to believe from sales people who only know what they've been told to say. You'll also learn that when it comes to electrical equipment, just like lumber sizes, nothing is quite what it says it is. For example, if you go to a lumber store and buy a two-by-four length of wood, you would think that it would measure exactly two inches by four inches. It doesn't. The same is true of nearly all standard lumber sizes. There are all kinds of reasons as to why this is the case, but the bottom line is...well, liar, liar pants on fire. The same is true for almost every single piece of solar power system equipment. Solar panels rated at 100 watts rarely ever exceed 80 watts even when angled directly at the sun. Inexpensive solar chargers rated at 30 amps rarely charge more than about 25 amps and almost never at full voltage necessary for a proper absorption charge on the batteries. So-called "smart" battery chargers are actually as dumb as rocks and utterly worthless. Inverters (devices that convert 12 volt DC battery power into 120 volt AC power similar to what you would get form a wall outlet in a house) never, ever, EVER put out the power rating listed on the box - Not even close! And batteries are the biggest lie of all with so-called "zero maintenance" batteries that you had better darn well maintain, charging specs that are almost always under-rated, amp-hour ratings that are nothing short of wishful thinking, and voltage output specs that are completely misleading.

Just to illustrate my point, you would think that a 12-volt battery would be fully charged if it read at resting (without a load) at 12 volts. But just like a two-by-four, you can't trust the numbers. A battery at 12 volts is closer to dead than fully charged. That's right, I said "dead". In fact, you should always strive to avoid allowing a deep cycle battery bank to discharge any lower than 12.2 volts. A fully charged 12-volt system with no load should measure at least 12.6 volts at rest!

To compound matters even worse, most information out there for solar power systems is written for grid-tie solar power home installations using very expensive equipment designed to connect to the power grid and function merely as a supplemental power source or short-term backup in cases power failure.

In fact, solar power systems come in three categories; 1) worthless, 2) well-researched designs by patient individuals, and 3) ridiculously expensive. It also doesn't help that sales people and retail companies that sell solar power equipment come in three similar categories; 1) worthless, 2) well-intentioned, but often lacking in real-world application, and 3) ridiculously greedy. Of course, there are some but few exceptions. I list them at the end of the book to help you get started without throwing you under the bus before school even starts.

So how the heck are you supposed to build a solar power system that meets your needs when there are so few reliable resources, people and information and virtually all of the equipment is labeled with misleading information? Well, buying this book was your best first step (thanks for that, by the way). The second step is to learn the triple-minimum rule of design.

Engineers know...wait, let me correct myself...GOOD engineers know that when they design a system, they over-design it so they can accommodate the unknown without starting a fire or causing an explosion due to component failure when the system is pushed to its stated limits. So, just to be doubly careful, when they list the specifications for their design, they round the numbers in their favor to allow for the occasional overextended application of their equipment. A perfect example of this is embodied by every engineer in Star Trek. During an emergency, the captain would ask for more than the ship could ordinarily do and the engineer would almost invariably admit that the specs could safely be pushed another 20% or so because they know what the ship can really handle. After all, they built the thing.

So now, back to the triple-minimum rule. When you decide to build you solar power system, you'll measure the collective power requirements of all of the equipment you plan to run on it. Whether it's just a single light, a laptop, and a small DC fan or it's a large screen TV, microwave oven and a crock pot, you'll have to add up all of the amperage (power current) values of the equipment you're likely to use at one time as well as estimate how long you might be using it. That may sound complicated, and it can be, if only because the rated power usage on device labels is almost always wrong - but that's the very reason for the triple-minimum rule. We'll also discuss ways to get way more accurate power usage information.

Say for example, you need a basic, low-power system to run a laptop, a light, a small DC electric blanket, and a fan. How do you figure out how much amperage each item uses? Unfortunately, yet again, you can't rely on what is printed on the power adapters or the back of each device. For example, my laptop power supply is rated at 60 amps, but when it's fully charged and I'm just checking emails or writing in my journal, it only actually pulls about 8 amps. That's a huge difference. Even when I'm watching a movie on my laptop and it's still charging the battery, it may only use as much as 50 amps at most. But if I'm doing video editing and conversion, it can pull as much as 100 amps! As for the light, you'll want to find a low-power LED lamp. They pull practically nothing in the way of power, but it is possible to get large LED lights that can draw enough current to make you take notice. And while the DC electric blanket will probably have the closest listed amperage rating, the fan could have three speeds with each one drawing more current than the next.

Now, you could estimate your power requirements by accepting the stated amperage values assuming they are all over-estimated like a good engineer would do. That is a viable solution, and admittedly a bit easier than measuring each item during load with a meter. Otherwise, you'll have to use a reliable method to measure the actual amperage being used by each item. To do that requires the use of a specialized electronic device called an ammeter and doing some basic math. We'll cover that later in more detail. Even more useful and reliable would be to use an inline battery monitor and then to test each item individually by simply plugging them in and taking note of the power usage values for each. But to do that, you have to already have a working battery power system installed, so that's kind of like putting the buggy in front of the horse.

Calculating the triple-minimum is actually pretty easy and you've probably already guessed how it works. All you have to do is estimate the amount of amperage you're likely to use at any given time from all of your available equipment and then estimate how long you might use each of them. Once you have these numbers figured out, you could simply double them to be safe, but instead, I recommend tripling them. Why? Three reasons: First off, ALL of your solar power equipment is conspiring against you by providing inflated, unrealistic specifications for functional capacity by as much as 20% or more for each and every piece. Second, all electronic equipment pulls more amperage when you first turn it on. Depending on the device, it can easily be more than twice the listed value. Particularly notorious for this are power tools and motor-driven devices. And third, if you require a fairly robust power system to run things like a microwave oven, toaster, hot plate or even an instant, single-cup coffee maker, you might forget that you're powering one high-amperage device and accidentally turn on another appliance requiring a high amount of power at the same time which when combined could easily push an under-estimated power design beyond its capabilities. So, if you don't just double your estimate but rather triple it, you have a far better chance of designing a system that can handle whatever you throw at it.

So, back to our example... Don't worry. All of this stuff will be explained later in more specific detail in later chapters. We have a laptop that may require 50 amps at most, an LED lamp that will need about 5 amps, a DC-powered electric blanket that pulls about 8 amps, and a three speed DC fan that pulls about 10 amps on high. That's a total of 73 amps. But you probably wouldn't run the electric blanket and the fan at the same time. And depending on how much you use your laptop and whether you have a reliable internet connection, your power requirements for your computer alone can fluctuate greatly. You might also turn the light off while watching movies on the laptop.

The reason you need to know both the amount of amperage used and the amount of time you might use each item is because of two things. First, power inverters are rated by the amount of power they can provide. Second, batteries are rated by the amount of current (amps) they can provide over time in hours. I know, this may sound kind of complicated. That's okay, as I said, we cover all of this is much greater detail later one easy step at a time. Don't worry, you'll get it. I'm telling you all of this ahead of time so you'll have a big picture of what you're in for later.

So, let's say you like to watch movies on your laptop with the light on while lying under your electric blanket from 5 pm until you fall asleep at 11 pm. That's six hours when the sun is no longer charging the batteries. So, you'll need a power system that can provide at least enough power to keep your laptop charged and running at full capacity to show movies along with your lamp and blanket. That comes to roughly 63 amps. Now, since inverters are rated in watts instead of amps, you have to multiply that value by the voltage which you would think should be 12 volts, but a better number would be 12.5 volts. 63 times 12.5 equals 787.5 watts. So you might think all you would need is at least an 800 watt inverter. Sadly, you would be wrong.

Even when an inverter is properly wired with short, heavy gauge wires as close to the battery bank as possible, a supposed 800-watt inverter may only reliably put out a little over half or two-thirds of that. Yes, that's right. It infuriates me, but there it is. So, when you triple your estimated power need (787.5 watts), you get 2,362.5 watts. Nearly 2500 watts?! Isn't that overkill? Yes, it is. That's the point. If you had tried to get away with a cheap 800-watt inverter, it very likely would not have worked AT ALL. So, if you doubled up and got the next available power spec inverter at 1200 or 1500 watts, you would be better off, but what if you decided to get a coffee warmer or a second lamp or a single-serve rice cooker? And did you forget to include the amperage required to charge and use your cellphone all day every day? That's another 5 to 15 amps. Just one more device can put you over the top on an under-estimated power system and you would have to start all over and buy a bigger inverter. And I haven't even mentioned your battery requirements yet.

**Power inverter.**

So, invest in a proper inverter. In this case, a good 3,000-watt inverter. Trust me, anything smaller and you're going to end up starting over and buying a bigger one. I started with a 100-watt inverter, then got an 800 watt, then a 1,500 watt and finally a 3,000 watt. Don't make the mistake I made and try to save money in the wrong places. Invest in a proper inverter.

You should also know that up to this point, I've only been talking about square wave inverters. Don't worry if you don't know what that means. We'll discuss this much more in a later chapter. Just know that they're cheaper and easier to find, but the AC electricity they provide is not like the power you would get from an electrical outlet in the wall of a house. For that you need a pure sine wave inverter. And those can easily cost three times as much as a square wave inverter of the same power rating. But you really only need a pure sine wave inverter if you're planning to run expensive equipment like a microwave oven, a home computer that's not a laptop, or a large screen TV. And not necessarily just because they're expensive, but because they were clearly designed to be used with normal house power from a wall outlet. Without a pure sine inverter, these devices will make disconcerting buzzing sounds and could even be damaged. More about that later.

So now, let's talk about batteries. This is where you need to know how long your power system is going to have to provide power. If you estimate that your total daily power usage will be roughly 3,000 watts. You'll have to convert those watts back to amps in order to determine the number of amp/hours you'll need from the batteries.

To do this, you divide the 3,000 watts by 12.5 volts to get 240 amps. (That's a lot by the way.) A good, affordable, deep cycle, battery for solar power systems is rated at 220 amp/hours. What this means, theoretically, is that after one hour of use at 240 amps on just one battery, you would have completely discharged the battery in less than one hour! But it's actually worse than that. Remember, every piece of equipment is conspiring against you. That battery will not be able to provide 220 amps in an hour. It will be less. When the battery is new, it will be slightly less. After a year or two, it will be noticeably so. Batteries get tired and they only last between 4 to 8 years at best with regular maintenance and proper charging.

So, you're going to need more than one battery. With this minimal power system, you should over-estimate to have at least two, preferably three, 12-volt deep cycle batteries connected in parallel (more about that later). Better yet, instead of 12 volt batteries, use six 6 volt batteries in series and parallel. (Again, don't worry about why I said that right now. Just remember there are reasons to consider other options. I'll give you hint though; 6 volt batteries have thicker plates which make them last longer.) With three 220 amp/hour batteries, you'll have roughly a 600 amp/hour battery bank. That will give you plenty of power when you need it and plenty of backup. Backup? Well sure. What if it rains one day and the sun doesn't come out at all to recharge your batteries? Worse yet, what if it's cloudy for two days in a row and then finally rains another two days? If you don't have plenty of reserve power, you won't last through the dark days when the sun doesn't shine.

Now I'm sure there are some of you who are yelling at me right now insisting that they know for sure that their buddy's RV or boat has just one little deep cycle battery and he's able to run lights and a little thirteen-inch color TV alllll night, so why in the world should you build such a huge power system? I mean come on, a big 3,000-watt inverter and three heavy deep cycle batteries? We haven't even covered how many solar panels we're going to need to charge all of that. Well, you're right, most smaller RVs, especially van conversions, have just one deep cycle battery. They're charged by a special, heavy-duty alternator on the engine with an isolator between the vehicle battery and the cabin battery - or from an onboard generator. But they're only good for very limited use and only for about a day; just enough for a day at the beach or a picnic. You need a power system that will be reliable and ready in spite of a couple of days without sunlight to recharge.

Of course, you can also hook up your battery bank to a generator, but that's another large expense - and as you'll learn later, completely unnecessary. And if you're thinking of connecting to your existing alternator on the engine to charge your batteries, think again. RV's have special heavy-duty, high amperage alternators installed. Modern vehicle alternators are rated just enough to do their job to keep the engine running and the headlights on and very little else. You might get away with connecting a second deep cycle battery in addition to the vehicle starter battery, but if that deep cycle battery gets drained too low, it will tax the alternator to failure trying to charge it back up. And even if you had a heavy-duty truck with a more powerful alternator that could handle the battery, if you used the inverter while it was connected to the battery and alternator, you would risk damaging the alternator by pulling too much current through the alternator.

**Battery isolator.**

If you insist on going this route, at least be sure to have an isolator installed so the alternator can charge both the vehicle battery and the cabin batteries simultaneously without one battery draining the other. The last thing you need is to have a long day out with the lights, fan and TV on in the vehicle only to find that you drained not only your deep cycle batteries but also your vehicle battery and now can't start the engine.

So, your goal must be to over-estimate and triple-minimum build your power system to accommodate excessive use and cloudy or rainy days. And what if you have a three-day weekend and you decide to watch movies all day? Whoops! You just tripled your power requirement. It'll be a good thing if you equally tripled your power capacity to be ready for it.

Okay, that was a lot to soak in at once. Again, don't worry. We'll cover every little detail in the chapters to follow. By now though, you should have an idea of just how complicated and difficult it can be trying to build your own mobile solar power system. But I guarantee you, after you read this book, it will all make much more sense. After you learn all of the stuff NOT to do, what's left is actually pretty easy and straight-forward. In fact, if all you learned was the following five items, you'd be better qualified than most solar power system sales people I've met:

- Proper panels
- Best batteries
- MPPT solar chargers
- Inverters
- Wiring

Forget about wall outlet powered battery chargers and generators. If you've built a proper solar power system, you will never need them - even to run a microwave oven, coffee maker, or a toaster oven!

So, let's get started!

Oh, by the way, if you don't know what a watt or an amp is or how to calculate either with volts, there's a chapter at the end of the book called Basic Electronics when you need it.

# Solar Panels

So why did I start with the first chapter on solar panels? The Introduction talked about inverters, solar charge controllers, batteries, wires and power usage, but nothing about solar panels. What gives?

Well, ironically enough, the solar panels are the easiest part of this whole endeavor. Once you learn what to avoid and what works best, there's really only one solution left. (This is actually true about solar charge controllers and generators too, but we'll get to those later.)

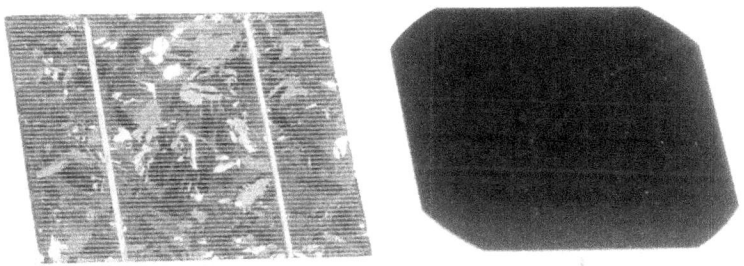

**Poly-crystalline and mono crystalline solar cells.**

Solar panels come in two forms; poly-crystalline and mono-crystalline. The difference is in how the solar cells themselves are manufactured. Poly-crystalline cells are blue-ish in color and when you look at them closely have cool reflective patterns of what looks like randomly layered flakes. They're actually kind of pretty. They're also the older method of making solar panels that dates back to 1954.

Mono-crystalline cells are the modern design we see most often today. They're a much darker colored cell with much less variation in appearance on the surface. They're also more efficient. So, as you might have already guessed, you want to focus on buying mono-crystalline solar panels. (At this point it would be more difficult to find good sized older poly-crystalline panels anyway, unless you found them on eBay or Craigslist. Just about every reputable solar panel company sells mono-crystalline panels exclusively these days.)

Solar panels also come in different sizes, usually rated by the number of watts they can supposedly produce. There are huge 250 watt panels, tall and slender 100 watt panels, and even square 25 and 50 watt panels and everything in between. This is good because the roof of your vehicle is undoubtedly going to force you to choose the size of panels that will fit while allowing you to optimize your total roof real estate. You can mix and match panel sizes as long as you remember one very important thing - you absolutely must have an MPPT solar charger (as opposed to a PWM model). Again, don't worry. We'll discuss this in great detail later. In fact, your choice of solar charger is THE most important decision you will have to make. And I'll warn you ahead of time. It's also the most expensive. Get ready for that.

**Renogy 100-watt mono-crystalline solar panels.**

Solar panels also come in two types of construction; 1) rigid aluminum or plastic frames with protective glass covering the solar cells, and 2) flexible sheets that can be molded to the shape of your vehicle. As of the writing of this book, flexible panels are far too expensive, unreliable and more easily damaged. So put them out of your mind. They're clever and even a little sexy, but that only helps slick sales people who are trying to make boats look cool. Trust me, forget them.

My preferred panel is the 100 watt Renogy. They're inexpensive, easily purchased and delivered from Amazon, and they're built locally in the U.S. Yes, you can find solar panels and kits at Harbor Freight all day for a lot less money. Trust me on this though, they're not the best choice. They're great for experimenting with the IDEA of having a solar power system, but the wires are too small for proper current delivery, the chargers are cheap PWM models, and the panels are insufficient to produce enough power for a fully useful mobile system. If you buy one, you quickly realize that you need many more solar cells, but once you do that you have to buy a new solar charge controller to handle the load. And all the wires have to be replaced with at least 6-gauge wire anyway, so don't bother.

You can also go big and buy a couple of huge 250 watt panels that are normally used for home installations. You just have to remember a couple of things. 1) With fewer, larger panels, you have less backup in case of damage or shade. If one panel is compromised, your power is instantly halved. And 2) You absolutely must have a more expensive MPPT solar charge controller to even begin to properly use the power from these larger panels. If all you have is a cheap PWM controller, you will only get the maximum power the charge controller can handle. And if the controller is too small, it could actually be damaged by the excessive power from the larger solar panels. (You should know however, that you should almost never buy a PWM charge controller anyway. More about that later.)

However, many solar panels you buy, it's best to try to make sure you cover every square inch of usable space on top of your vehicle with solar cells by using various sized panels to fit wherever you can. It's a simple concept, but it bears repeating; there's no such thing as too much solar power. Sure, you might get away with less if you live in the Arizona desert, but try spending a week in Oregon or

Washington State in the rainy season and you'll wish you had a trainload of panels hitched to the back of your vehicle.

And even if your power needs are slim or you do happen to live in a place with more yearly sunlight hours, having more active solar panels will charge your batteries quicker in cases when you might need more power some days more than others. Case in point: I have a swamp cooler in my truck that draws about 25 amps. In one day of full use on a particularly hot day here in San Diego, my batteries can easily drop below 80% before sundown just to get the internal temperature of the truck below 80°. A couple of days like that without adequate solar generation from a full roof of panels and you'll be in trouble soon. But since I have my entire roof covered with solar cells, my batteries are often fully recharged sometime around lunchtime. But it wasn't always that way.

When I first got started, I bought two of the huge 250-watt grid-tie house type solar panels and connected it to a 60 amp PWM solar charge controller. It worked - sort of. But because the PWM controller could not make full use of the power coming from the larger solar panels, I was only getting about half of the power to the batteries than I should have. The rest of the power was effectively ignored by the charge controller. I'll explain why later, but I'll give you a hint as to why right now.

MPPT stands for "Maximum Power Point Tracker". PWM stands for "Pulse-Width Modulation". So here's the short explanation. PWM is the basic method of converting solar power into battery power. It's not too smart - voltage in, voltage out. MMPT uses the right kind of PWM at the right time to make sure your batteries are charged properly with the right voltage depending on the amount of sunlight, the current charge on the batteries and a schedule for desulfation (usually referred to as "Equalization"). I know, right?! What? Don't worry, it will all make sense later when we talk about solar chargers.

So, as you might have guessed by now, I had two massive solar panels just screaming power at a deaf PWM charger and my batteries suffered because of it. Not only because the PWM controller couldn't make the best of the power it was being offered, but also because PWM controllers (even the pricier ones) are notoriously designed and programmed improperly and will never fully charge your batteries. Trust me, I had 500 watts on the roof and there were many nights I went without even a light to read a book! San Diego has a weird weather pattern in the Spring we call "May Gray" followed by "June Gloom" during which for two whole months you're lucky to see the sun peak from behind a dense layer of gray clouds for more than a day or two at a time. This, in combination with a useless PWM charge controller equated to dead batteries.

Later, out of just shear ignorance, I figured I must need more solar panels if I'm going to have any chance at lasting out the gray nights. But I had learned a bit since I last bought solar panels, so I decided I needed to buy smaller panels and more of them. I was also running out of space on my roof pretty quickly.

One lesson I learned was that even a small shadow over just a fraction of a solar panel will reduce its power output dramatically. And with just two big panels, if one was in the shade or the movement of the sun caused a shadow to move over the panels later in the day, the power getting to the batteries was minimal at best.

So, I bought three 100 watt solar panels to add to the two big ones I already had installed. These panels were tall and slender and three of them side to side measured about the same size as one of the larger 250 watt panels. I was also later able to add another three 100 watt panels laid end to end to cover the rest of the side of the roof. I eventually had 1,100 potential watts of power on my roof! But my poor PWM charger was only able to use about HALF of it at best. But I learned. Trust me, MPPT. It's the most expensive item you'll have to buy, but you really don't have a choice.

## Panel Placement

POINT THEM TOWARD THE SUN! I know that sounds like a ridiculous statement, but if you look around you'll see entire homes covered with solar panels aimed at nothing. What were they thinking? Of course, you have little choice with the roof of a vehicle since it's just a flat surface, but as it turns out, flat on the roof is better than pointing in the wrong direction.

Everyone knows that the sun rises in the East and sets in the West. So, which direction do you point your solar panels? Well, if they can be automatically angled and rotated, then obviously they will be pointed directly at the sun at all times with mechanical motors - no worries there. But most times, having that kind of motorized equipment just isn't feasible or even allowed in some housing districts. So most residential solar panels are mounted flat on the roof of the house. And of course, most houses were not built with solar panels in mind. Consequently, what you end up with is panels mounted on the roof facing toward the East or West in a vain attempt to try to capture either the rising or setting sun. This is pointless.

The light from the sun is worthless when obscured by morning haze or evening clouds. The best condition for solar power generation is direct sunlight on a cool day. And the only way to mount solar panels flat on an angled roof that makes any sense is if the panels are facing the equator. In the United States, that means toward the South. As the sun rises in the East and begins to rise above the morning haze and fog, the panels at least have a chance at a glancing blow of sunlight which improves as the day progresses. At high noon, the sunlight is directed straight into the panels for maximum power generation. Then as the sun begins to set, the panels still have a chance at the last bit of light before the sun dips below the horizon.

Panels mounted flat on the roof of a vehicle have about as much of a chance to collect solar energy as a decent home installation angled to the South. No, it's not ideal, but it is very functional. When you factor in the morning and evening haze along with the extreme angle of sunlight, the early morning and late evening hours are a wash. But having the panels pointing straight up during the high sunlight hours is almost optimal - well, as optimal as you're going to get. So, it balances out. Besides, you want to make sure your panels are securely mounted to the roof. Any kind of method to allow angling the panels toward the sun will undoubtedly be less secure. Remember, these panels have to stay connected to the top of a vehicle doing 65 miles per hour on the freeway!

## Mounting

Most solar panels are mounted to the roof of a house with standard mounting brackets. But when mounting panels to a mobile vehicle, you have to get a little more creative and be a lot more attentive to safety while also being sure to avoid creating leaks in the roof of the vehicle.

I've seen all kinds of creative solutions. Some have used ski racks mounted on the roof using the straps and braces that came with it and then bolted the solar panels onto the rack. I've seen people do something similar on vans and trucks using standard ladder racks as a base for the panels. I've even seen an installation where they literally glued the panels to the roof without using any brackets at all. Of all the mounting methods I've come across, only one was tragically misguided because they used straps across the panels to hold them down. The straps put a permanent shadow over the areas of the panel where they overlapped and power generation was severely diminished because of it.

With my truck, I had a huge, flat, metal surface to work with. I drilled small holes on the side of the panels and into the roof, then mounted the panels with simple, inexpensive "L" brackets you can find in any hardware store or big department store. To mount the "L" brackets to the panel and the roof, I used flaring molly bolts that spread out on the other side of the hole when you screw them down. In order to try to prevent any leaks in the roof, I put down a dab of weather-proof sealant around the hole before setting the molly bolt. Sadly, this almost worked. I later had to spray leak preventing goop on the ceiling from the inside and still never got all of the leaks. The problem is motion. If there's any motion in the mountings, leaks will eventually occur. The roof of my truck is just one big sheet of metal that even before I started poking holes into it was prone to temperature expansion and contraction, vibration and movement from the wind racing over the top at highway speeds.

So, as you can see, it's not quite a straight-forward endeavor getting solar panels mounted to the top of a vehicle. The biggest issue is wind. It's imperative that the panels be mounted sturdily enough to withstand the strong winds that will whip over the top of a vehicle when driving on the freeway at 65 miles per hour. Luckily, my "L" bracket solution kept the panels flush to the roof and held them down tightly so I never had an issue (other than a couple of leaks).

So, regardless of which method you use to mount your solar panels, it's vitally important to make sure they will stay mounted without causing leaks. The best option appears to be the ladder rack or luggage rack approach. This allows you to secure the panels firmly to the rack instead of the roof, avoiding any chance of leaks. But it also stands out clearly for everyone to see that you have solar panels mounted on the top of your vehicle. If this isn't a concern, then no worries, but in my case, I was striving for a more "urban camouflage" solution where my 14-foot box truck would continue to look like nothing more than a standard delivery truck without having solar panels sticking up on the top.

Whichever method you decide to use to mount the solar panels, just keep in mind the strong winds at freeway speeds, potential for leakage in the roof, and safety from theft. If someone can walk up to your vehicle and easily reach the mouthing brackets, try to make sure the screws or bolts are secured somehow or even stripped out if you need a cheap solution.

## Series, Parallel or Both?

Each solar panel generates a specific amount of maximum voltage and current depending on the size and type. But all of them function at a baseline voltage isolated from each other so they can be connected together to form a large array of panels without damaging each other.

Just like batteries, solar panels can be connected in series or in parallel. Each option affects the power output differently. In series, the voltage increases, but the current the stays the same. In parallel, it's the exact opposite. This is convenient when deciding what baseline voltage system to design for a residential installation because you have options for 12, 24, or 48 volt systems. But for mobile systems, the larger 24 and 48-volt systems are expensive and way overkill. So, for mobile solar power systems designed for 12-volt equipment, the question is kind of moot; parallel - period. End of discussion.

That being said, there's always somebody out there who thinks they can outsmart themselves and be cleverer than the next guy. They logically deduce that higher voltages don't require as heavy of wiring, so by connecting the panels in series or a combination of series and parallel to get higher voltages, they can theoretically save money on wiring. The problem with this logic is it's nonsense. There's no appreciable difference between 12 volts and 48 volts when it comes to wire size requirements. And it's really easy to show why. Whichever way you connect solar panels, the same amount of power in watts will be flowing down the wires. Amps times volts equals watts. And if you put too many watts through a thin wire, you're going to lose power just to heat up the wires like a light bulb filament.

This all assumes you're using an MPPT solar charger. PWM chargers can rarely handle higher voltages, and even if they could, they can't convert between current and volts when needed for proper efficient charging. But even for an MPPT charger, if you connect it to higher voltages from the solar panels, it will simply convert the excess voltage amps for whatever it needs at the time. It's a moot issue. For mobile systems, parallel is the only way to go.

One other key thing to keep in mind is that when you connect solar panels in series, if one of the panels is compromised by a shadow, the other panels in the same series are compromised. This isn't the case with a parallel installation.

## Panel Maintenance

Depending on where you live and the kind of weather conditions you experience, solar panel maintenance can be something you do once every three to six months or every two weeks in the extreme. A lot of it depends on how much rain and dust you have to deal with. In areas that get a lot of rain, but have little dust (or pollen), the panels will remain fairly free of dirt for a long while. The same goes for very dry areas that may have a lot of dust and sand but have little moisture.

There are two things that can severely limit the power generation capacity of solar panels; dirt film from dust trapped by moisture, or shadows from trees, bushes, fallen leaves or even antennas. One of the worst environments for solar panels is near the ocean in dusty or pollen-covered areas. A solar panel can very quickly accumulate a dirty film from mild moisture and dust that collects in the morning and is then dried hard onto the surface by the sun.

Of course, this accumulation of dirt is fairly easily solved by simply cleaning the panels on a regular basis. For some, that means crawling up on the roof about twice a year and either spraying them down with a high-pressure hose or, better yet, using a mop to clean them with window cleaner. For others, it can be a monthly chore because of all the dirt and moisture in the air. With my truck, living near the ocean in San Diego and San Francisco, I would easily have to check the panels every three month during the cooler season.

The only other maintenance required is to check the wires and mounts to make sure there are no frayed or cracking cables that could short out or mounts that have been loosened by wind or rot, especially on a vehicle that may be traveling at freeway speeds often. The last thing you want is for a solar panel to loosen up from the roof of the vehicle, catch the wind and fly off on the freeway.

## Solar Panels Recap

Buy only mono-crystalline solar panels. Buy as many as will fit on the roof of your vehicle of whatever size panels will fit AS LONG AS you have a proper MPPT solar charge controller.

Wire them in parallel to maximize available current and avoid compromising your panel array with shadows.

Don't worry about trying to angle the panels toward the sun.

Be sure to securely mount the panels on the roof to be able to withstand freeway speeds.

And clean your panels as often as necessary depending on the climate and dust in your area.

There. Okay. Simple so far. Now let's talk batteries.

# Batteries

I could go on and on about the technical aspects of different types of deep cycle batteries and drone on about each type, but in the end your choice of batteries comes down to just two issues; cost and convenience.

But first, it is important that you know to buy only deep cycle batteries. Some "marine" type batteries are hybrid types to be used with trolling motors and are not suitable for solar power systems. If the words "hybrid", "cranking", "starter", or "dual-purpose" are mentioned anywhere, it's the wrong kind of battery. "Golf cart" or "forklift" batteries are usually okay though. In fact, they're usually the best choice because they're designed for deep discharge, quick recharge and long use.

Also, today's solar power batteries don't have a "memory" like the old Nickel-Cadmium type batteries used in older cell phones or wireless devices. So there's no need to completely discharge the batteries and then recharge. Modern solar power batteries are designed to be discharged and recharged several hundred times during their lifespan and can accommodate a deep discharge as easily as a short one.

## Flooded Lead Acid Batteries

That said, the least expensive battery option is flooded lead acid batteries. They are the least expensive option, but you do have to maintain them on a regular basis. The liquid inside is a combination of acid and distilled water. The water will "boil off" with regular charging over time, requiring you to periodically replenish the water with distilled water. During a high voltage desulfation/equalization charge, noticeable bubbles will appear inside the cells. This "boiling" causes the water to evaporate leaving the acid behind and must be replenished on a regular basis. (Be sure to use only distilled water, not tap water or drinking water. The minerals in the water will shorten the life of the battery.)

**Trojan 6 volt flooded lead acid battery.**

To replenish the water in flooded acid batteries, you simply pop the cap off the top and slowly pour fresh distilled water into each cell. It is very important that you do not overfill the cells or splash the acid water out of the cells while filling! Take it slow. Leave enough room in the cells for the water to expand when the batteries heat up during charge. The water should be just above the plates and never be allowed to drop below them exposing them to air.

The so called "boiling" effect isn't really boiling like you would get with a tea kettle. The bubbles are actually caused by a chemical reaction by the introduction of electricity into the acid/water. The electricity causes a process called electrolysis which causes the hydrogen and oxygen atoms to separate in water. When this occurs, the batteries outgas the hydrogen and the water evaporates. It's a normal function of the design. That being said however, batteries should never get hot while charging. Warm is okay and expected. You should be able to comfortably touch the side of the battery without any concerns of burning or extreme discomfort. Also, it's perfectly natural for batteries to swell a bit with time, but if they bulge to the point of wobbling, you've got a problem.

This outgassing of hydrogen is a bit of an issue on a couple of points as well. The area where you store the batteries needs at least some minimal ventilation to outside air. This can be tricky with a vehicle installation. Hydrogen is explosive. HOWEVER, there's no need to panic. The amount of hydrogen released is minuscule. Worst case scenario, in a completely air-tight space for several sunny days of full charging or a high voltage desulfation charge, might result in a noticeable level of hydrogen in the surrounding air. Hydrogen is odorless, but it is noticeable by the lack of breathable air in the same space. It's highly unlikely however, that a vehicle would be that air tight. Regardless, this is an issue. But it's actually an issue for a completely different and unexpected reason; corrosion.

**Battery terminal corrosion due to hydrogen outgassing.**

Hydrogen in the air will bind with metals and start corroding wires and battery terminals very quickly. Without proper ventilation, this is a more irritating problem. The terminals and wires will first turn a bluish-green color and then build up a kind of cottage cheese nasty wet powder that if not cleaned immediately can harden making it even more difficult to clean.

**Corrosion prevention spray.**

The good news is you can easily prevent this. There's a special chemical you can buy to simply spray onto the clean terminals and exposed wires to prevent this chemical corrosion from occurring. You can buy it any auto parts store. It has a thin red tint and can get a little sticky. I've also heard of people using WD-40 instead. But with that you trade sticky for oily and it doesn't last near as long. So, use whatever works for you. The WD-40 is a cheaper option, but you have to keep applying it on a regular basis.

With all of these drawbacks; outgassing, dealing with acid, having to replenish water levels, and corrosion, you could easily be excused for wanting to move on to another battery type option in spite of the cost savings. Luckily, if you have the money, you have two more options.

## AGM or Gel Cell Batteries

The next type of battery you might choose is the AGM or Gel Cell battery. These types of batteries don't have to be monitored for water levels and they can even be placed on their sides if need be, unlike flooded lead acid batteries that must be kept upright to avoid leakage. These two things make AGM batteries more convenient, but there are a couple of drawbacks. AGM batteries don't hold a charge as long as lead acid batteries and they cost a lot more. So you'll have to buy more batteries at a higher cost than you would with lead acid batteries. You'll have to consider the pros and cons on this one for yourself.

**AGM battery.**

Personally, I opted for the far less expensive lead acid batteries. Sure, about once a month I have to check water levels and the battery terminals are covered with a thin sticky layer of red anti-corrosive spray, but it's really no big deal. I saved a heck of a lot of money and, quite frankly, there aren't too many places to cut corners with a solar power system. Also, when AGM batteries start to go bad, there's nothing you can do about it like you can with lead acid batteries. While I don't recommend it, you can resuscitate an aging lead acid battery which can give you another six to eight months of life while you locate some new batteries. You can read more about his in the chapter on Battery Resuscitation Tricks later in the book. With AGM you would gradually lose capacity until one day you just wouldn't have any lights.

## Lithium Batteries

The third and final option for batteries is lithium. They hold a good charge, but even better they recharge quicker than the other battery types. They don't require any maintenance either, so they're very clean. But there are two huge problems; they're insanely expensive for the amount of capacity you get and there's a slight issue of a potential hazard of explosion in extreme cases of heat or puncture.

I would love to be able to afford a lithium battery bank, but it would require twice as much storage, four times as much money, and a hundred times more diligence in cases of fire or collision. So, nope! They sure are pretty though.

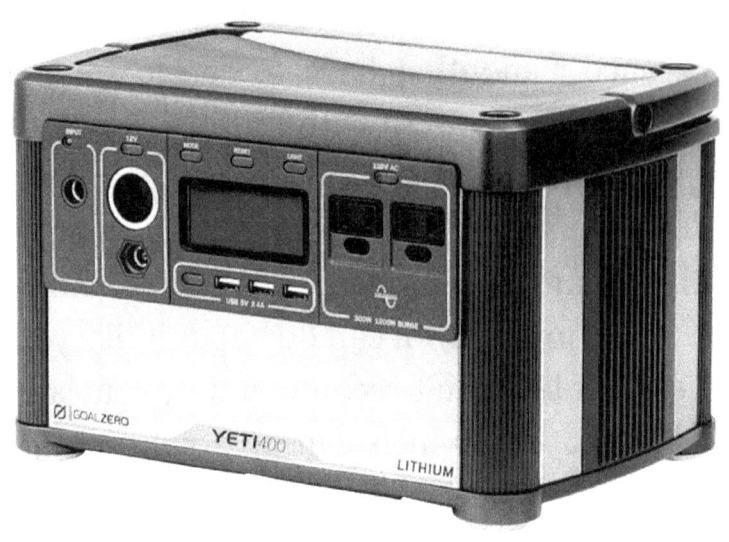

**Lithium battery system.**

One other extremely important thing to keep in mind about lithium batteries is (and this is for you clever electrical hobbyists out there) you can't just buy lithium batteries and connect them like any ordinary battery. That's why they are built into full blown systems like the one pictured above with built-in charging mechanisms and monitors. The reason for this is very simple. Lithium batteries are not designed to be fully discharged. If you drain a lithium battery until it's dead, that's it. There's no coming back. They're dead for good. Most lithium power systems are designed to allow the internal batteries to discharge to only slightly below 50%, but register as fully discharged. Some of you might think you're clever enough to avoid discharging lithium batteries that low without a proper charging and monitoring system by just keeping an eye on the charge, and you might be right...until the one time you're not. Then all you have is really expensive paper weights.

So, bottom line, it comes down to flooded lead acid or AGM. And the only reason you should consider AGM is if your vehicle installation is just too cramped for easy maintenance or you really, really don't want to be bothered with an occasional water replenishment routine and you're willing to sacrifice more money and reduced charging capacity. I make no judgments either way. I've seen some spectacular AGM installations. They needed more batteries than they would have needed otherwise with lead acid, but they never had to look at their batteries again - and they could afford it, so there's that.

## 12 Volt or 6 Volt?

This is an easy question, so this will be short and to the point. 6-volt deep cycle batteries have larger lead plates than the 12 volt batteries which allows them to hold a deeper charge. Consequently, they provide a longer charge. But more importantly, as they age and the plates decrease in thickness due to unavoidable sulfation build-up, they will last longer than the thinner plates in the 12 volt batteries.

So, buy 6 volt deep cycle "golf cart" batteries and wire them in series for 12 volts. For more amp-hours, create a battery bank by wiring identical two-battery series sets and then wire those sets in parallel with the previous set. Check out the Basic Electronics and Example System Diagrams toward the end of the book for details.

This is true for lead acid and AGM type batteries. Lithium battery systems are a bit more complex and have their own interconnectivity options.

And what about 24-volt and 48-volt systems? Forget it. These higher voltage systems have their place; residential installations. The equipment is far more expensive and frankly, unnecessary for mobile solar power systems.

## Mounting and Placement

Allow me to get one myth out of the way right off the bat. Batteries will not discharge any faster because they're sitting on a concrete floor as opposed to a wooden or metal shelf. I don't know where this nonsense came from, but it's completely false. Forget about it.

Now then, mounting methods for solar power batteries are different for each of the three types of batteries. Flooded lead acid batteries require the most attention. These types of batteries must be kept upright and secured from tipping over in order to avoid spilling the acid inside. AGM or Gel batteries on the other hand are sealed tight and can be mounted on their sides without any worries. I've even seen them mounted upside down, but I don't recommend it. It's just weird.

Lithium battery systems are enclosed in cases for protection and are often too big to do much else with other than let them sit where they are, but it is very important to know one thing; unlike lead acid and AGM batteries, lithium batteries can be a serious concern for explosion or fire if exposed to excessive heat or damage leading to puncture. Consequently, the manufacturers have built sturdy enclosures and some even include cooling fans just in case. But if you're involved in a serious collision, all bets are off. This assumes you opted for a professionally manufactured lithium battery power system and didn't instead try to build your own with individual lithium batteries. Don't try that by the way (unless you're a professional electronics engineer) because unprotected lithium batteries can be dangerous, but more importantly lithium batteries require a complex charging system that never allows the batteries to completely discharge because once that happens, the batteries are irretrievably dead. They will not work ever again.

Usually, the toughest part of mounting your batteries is finding a place in your vehicle where they'll fit. And of course with lead acid batteries, there's also the case of proper ventilation. Truck campers, van conversions and RVs usually have a designated space already built into the vehicle that allow for one or two batteries. Unfortunately, for higher capacity power systems requiring additional batteries, you'll have to find the space yourself.

My solution was to remove the passenger seat in the front of the truck, cut and sand the chair mounting bolts sticking up from the floorboard to avoid damaging the batteries, and then built my lead acid battery bank right on the floor. The truck floor was made of a nice hard rubber material, so it provided the perfect combination of traction and protective surface in case of spills for easy cleanup. In fact, I once just took the truck to the car wash and power sprayed the floorboard after a hard-learned lesson about overfilling the batteries with water, spilling acid all over the place. Ah, such fun memories. The batteries were heavy enough to avoid any major shifting while the vehicle was in motion and the heavy 4-gauge interconnecting wires I used to connect the 6 volt batteries in series and parallel helped hold them in place as well. I considered building some kind of bracing frame for the battery bank, but it was such a snug fit, I didn't see any reason for the extra effort.

With AGM batteries, you have a little more leeway on how you mount them since you can safely put them on their sides without fear of spillage. But they still take up a lot of space. Since they don't have an outgassing issue either, you can be a little more creative about where you store them as well since they don't require the kind of ventilation lead acid batteries do.

Lithium battery systems are often bulky cases with lots of pretty switches and displays that you just have to find a spot for that's not in the way. I've seen them shoved under beds or even used as a makeshift table. Some models even have wheels on them, so you can roll them around. They don't need much ventilation like lead acid batteries, but they should be afforded at least some breathing space in case the batteries get hot in an enclosed vehicle.

Regardless of your battery choice, it's very important to consider two things; heat and collision. Batteries should be stored away from direct sunlight and allowed to breath as much as possible if the vehicle will be exposed to extreme heat. As for collisions, well…they will happen. That's why they call them accidents. It might not be your fault either. But no matter who might be at fault, a direct collision with a bank of acid-filled batteries can be very dangerous. And a compromised lithium battery system can actually explode and/or catch fire. So, whatever you can do to secure your batteries, do so with extreme diligence. This may be one of the stronger selling points for sealed AGM batteries. They might be the safest option in case of a severe accident.

I had an acquaintance who had more money than brains and owned three very expensive mobile lithium power units. They each had wheels and could easily be rolled around his 40-foot RV to provide power wherever he needed it. In spite of my warnings, he refused to tie them down or secure them in any way. He felt the friction from the carpet and the heavy weight of the lithium battery units would keep them in place well enough. Sadly, he was involved in a glancing head-on collision on a two-lane highway in rural Texas and the abrupt decrease in speed caused all three of the units to rocket to the front of the vehicle. One of them pinned and broke his right leg in two places. At eighty-two years of age, he still hasn't fully recovered after two years and has to walk with a cane now. He now has two of the lithium units mounted firmly under his bed in the back. The third one was damaged and burned beyond repair.

Another good friend of mine actually set his two 12-volt batteries on the large flat dashboard of his older model RV. I begged him to move them, but he insisted they had been like that for years and he never had a problem. Not a month after that conversation, he had a mild heart attack, lost consciousness while driving, and ran into the back of a line of cars. Both batteries shot right through the windshield and landed in the back of the car in front of him. Thankfully, the baby seat in the back seat of the car was not occupied at the time when one of those heavy batteries shattered the back window and rested squarely in the seat where the baby would have been.

Okay, sad stories. Don't freak out. Just let them serve as valuable lessons. Batteries are heavy and can be dangerous, if not downright deadly, if you don't take the time to maintain and secure them properly.

# Wiring Configurations

## Series Battery Connection

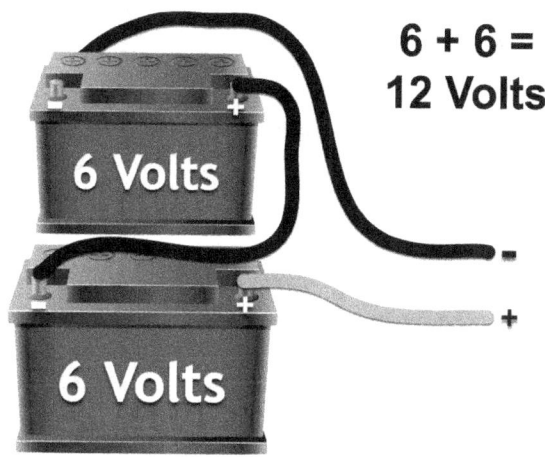

**Series battery connection.**

Connecting two batteries in series is very simple. Just connect the positive terminal of one battery to the negative terminal of another and then use to the two remaining terminals as if both batteries were one big single battery.

## Parallel Battery Connection

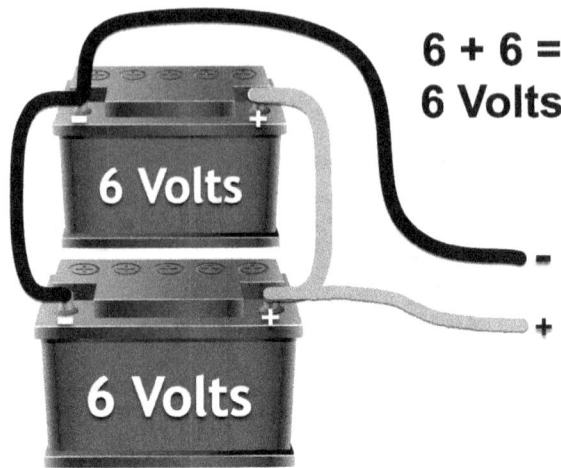

**Parallel battery connection.**

Connecting two batteries in parallel is just as simple. Connect the positive terminals from both batteries together. Do the same for the negative terminals. Then connect the two furthest opposite terminals as if both batteries were one big single battery.

## Series and Parallel Battery Connection

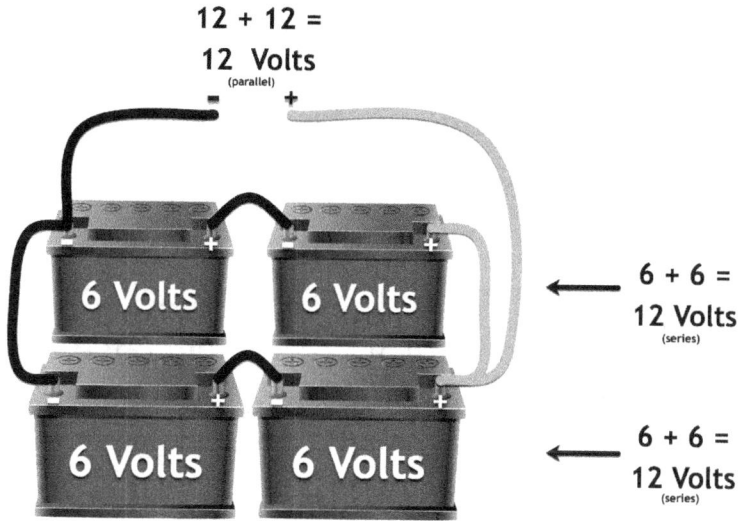

Series and parallel battery bank connection.

Connecting four batteries in series and parallel just combines both methods. He important thing to always remember with multiple battery banks is to always connect to the further opposite terminals so you are sure to use all of the batteries equally.

# Battery Charge Expectations

First off, it's important to know that I will only be discussing 12-volt power systems. 24 volt and 48 volt systems are best used for residential installations and are far too expensive and overkill for any mobile solar power system, no matter how large.

I've only seen one RV in my lifetime completely decked out with a massive solar panel array with auto-tracking panel mounts, a huge lithium battery bank in its own special ventilated and indirectly lit closet, dual parallel solar chargers, front and back power monitors, double-redundant 5,000-watt pure sine inverters, two interconnected backup generators, and a dedicated monitoring laptop on an RV bigger than my high school that was all based on a 48 volt system. They said it cost nearly $80,000 for equipment and installation. And I guarantee you, they never even walked past the isle in the book store where this book is for sale. They basically built a residential grid-tie solar power system on top of a 45 foot RV. But as it turned out, it didn't work worth a crap because of the wiring and he had to run the generator every night. You can read more about that later in the chapter on Wiring Types, Gauges and Length.

My point is you don't have to spend a ridiculous amount of money on massive amounts of equipment requiring 24 or 48 volts that were designed for residential installations for you to have a robust and very functional mobile power system using far less expensive equipment designed for 12 volts.

That being said, it is vitally important that you know what kind of voltages you'll really be dealing with in a so-called 12-volt system.

First off, if your battery is at 12 volts, it's dead! You read that right. Crazy huh?

It's bad enough that you can't trust the information on manufacturer labels for power usage on electronic equipment and power capacity for inverters, but battery voltages are the biggest lie ever. A 12-volt battery (or two 6 volt batteries wired in series) should never drop below 12.2 volts. And charging voltages are much higher than you might expect.

You should also know that charging a battery is not just attaching wires and throwing volts and amps at it until it's full. Batteries must be charged in very specific stages or modes with predetermined voltages and current levels to properly and fully charge a battery without damaging it.

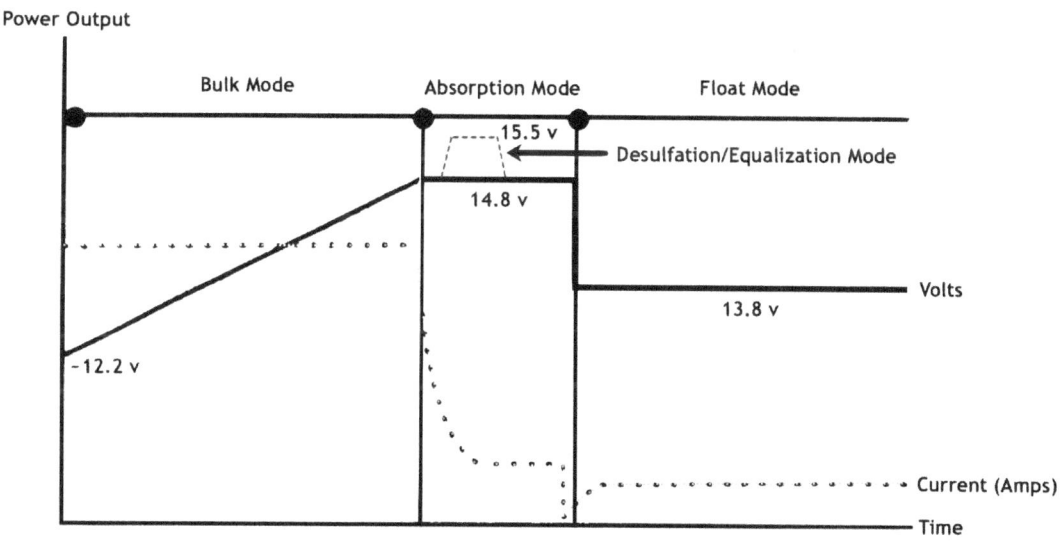

**Charging modes.**

The first stage is called Bulk mode. Doing this stage, the voltage slowly ramps up from the discharged state voltage of the battery to a maximum of about 14.8 volts. The starting voltage depends on how much the battery was discharged to begin with. If it was nearly fully discharged, it could be as low as 12.4 to 12.6 volts. During this bulk charge stage, the current is constant. Essentially, the charger is feeding the battery as much power as it can safely consume. You can't just throw 15.5 volts at 80 amps at a battery and hope it will charge faster. It simply won't take any more power than it can. That's physics. Actually, it's chemistry, but you get what I mean. All it will do is heat up and possibly damage the battery.

Once a batter reaches a charge that allows the full charging voltage of 14.8 volts, the charger will change to the next stage; the absorption mode. In this mode, the voltage is kept at a constant 14.8 volts while the current is gradually decreased as the battery nears full charge. Again, you can't force this with higher current. The battery will take only as much as it can safely manage over time. And it gets harder and harder to fully charge a battery as it approaches full charge.

Once the battery has absorbed as much power as it can fit, the charger changes into the final stage; float mode. Float mode is a maintenance mode. It keeps the voltage at a safe 13.8 volts and low amperage just to keep the battery topped off.

There is one other charging mode called Desulfation or Equalization. This mode is only used about once a month to make sure that the battery not only gets a solid, full charge, but also gets a high enough voltage to knock loose any sulfation deposits that may have built up on the cell plates. Sulfation is caused when a battery is not fully charged over an extended period of time. Sulfation kills batteries. It is a crystalline substance made of combined lead and sulfuric acid to form a light purple powdery substance that builds up on the plates in the cells. The desulfation mode will occur the moment the absorption voltage (14.8 volts) is reached. At that point, the charger increases the voltage to 15.5 volts for about three to four hours depending on the program settings on the charger. Four hours tops is enough.

Once the desulfation charge is completed, the charger will drop back into absorption mode briefly until it can reliably drop into float mode.

As you can see, these are much higher voltages than 12 volts. But it makes sense, batteries are designed to allow electricity to flow out of them. If you want to put electricity into them, you have to "fight the stream" at a higher voltage.

Now then, you already know that a 12-volt battery should never dip below 12.2 volts and 12 volts is actually dead. That's misleading enough as it is. But you should also know that a fully charged battery at rest with nothing running off of it, should register at least 12.7 volts. Basically, that means that 12 volt batteries are really 12.7 volt batteries.

Knowing these voltages is extremely important, and for some reason hard to come by unless you look for them. Consequently, salespeople who sell inverters and solar chargers, often don't know enough about batteries and the required voltages so they don't know when their equipment is providing too little power or pulling too much. For example, many inexpensive PWM solar charge controllers rarely charge above 13.8 volts. They never reach proper absorption voltages. This leaves the battery at 80% of charge at best. Then to compound matters, inverters are often programmed to shut off if the battery voltage drops below 10.5 volts. So, most sales people think that a battery at that low of voltage is an acceptable scenario. It isn't. EVER!

If your battery voltage is dipping into the 11 volt or 10 volt range, you're doing something terribly wrong. Your inverter should NEVER shut off due to low voltage. Please be clear on this. A 12-volt battery is really a 12.7 volt battery and should never drop below 12.2 because 12 volts is dead. Sure, the inverter will continue to work until the battery drops down to 10.5 volts, but by that point, it would take multiple days at full sunlight to fully recharge your batteries. And every second a battery is not at full charge, it is building up sulfation crystals. If a battery never gets fully charged, the sulfation will eventually short out the plates in the cell and kill it permanently.

There is one and only one exception to this rule about 12.2 volts. When you use an extremely high-powered appliance like a coffee maker, microwave or toaster, the voltage can temporarily drop to as low as roughly 11.4 volts. But as soon as you turn off that appliance, the battery should instantly register at its original 12+ volts. This is more a function of the battery monitor than the actual voltage in the battery. The monitor is watching how much power passes through it, not what is actually in the battery. An appliance pulling that much power doesn't give the battery monitor much to work with while it's using all that power. That's the short answer. Technically, the resistance inside the battery increases dramatically while a heavy current is being delivered and battery monitors use resistance to calculate voltage. This momentary increased resistance provides misleading information to the monitor while heavy current in being pulled from the battery. But once the heavy power drain ends, the monitor will show the true battery voltage once again.

The other aspect of battery charge that is vital for you to understand is amp hours. The term "Amp hours" describes a total value of electrical power measured in amps that a battery has stored which will last for a specified number of hours. Deep cycle batteries are rated in amp hours as opposed to cranking amps for regular car batteries. They're actually the same kind of measurement, but batteries measured in cranking amps are designed to deliver high amps for a short period of time while deep cycle batteries measured in amp hours are designed to deliver that power over a much longer period of time.

It makes sense to have two different kinds of batteries in this way. If you need to start an engine, you need a lot of power all at once, but if you want to run lights and a fan all day, you need a battery designed to deliver that power more slowly.

A typical flooded lead acid battery amp hour rating is about 220 amp hours. This means for example that, in theory, if the battery is fully charged you should be able to run a 20 amp electrical device on that single battery for 11 hours before it is completely discharged; 220 amp hours / 20 amps = 11 hours. Pretty simple, right? Except, as always with solar power equipment, there's a catch.

As the battery discharges during use, the amount of current in the battery decreases, but so does the voltage. And unfortunately, the voltage will drop below a useable level before the full amount of amp hours in the battery is reached. So even though you may think you might have 11 hours for the fan and lights, you might only actually have enough for about 8 hours before the lights start to dim.

This is yet another reason to triple your power needs assessment instead of doubling. You can read more about that in the chapter on Determining Power Requirements later in the book.

The good news is, you can increase the total amp hours you have available by creating a larger battery bank. When you wire batteries in series, you double the voltage, but when you wire them in parallel, you double the amp hours. Pretty cool, huh?

So, for example, with my battery bank I have six 6 volt batteries. They're wired in three parallel groups of two batteries in series. I first wired two 6 volt batteries in series to double the voltage to 12 volts. To do this, you simply connect the negative terminal of one battery to the positive terminal of the other battery. Then when you measure the voltage across the unconnected terminals of each battery, you get 12 volts. I did this two more times to create two more sets of double 6 volt batteries that would each provide 12 volts. Then I wired those three sets of double batteries as if the two batteries were one big battery each. In other words, it was just like wiring three big 12 volt batteries in parallel. To do this, you simply wire the same outside (unconnected) terminals of each of the two batteries because the two inside terminals are already wired together for the series connection. So, all three positive terminals are wired together and all three negative terminals are wired together.

By wiring the batteries both in series and parallel, I doubled the voltage to 12 volts, but I tripled the amp hours to 660 (theoretical) amp hours.

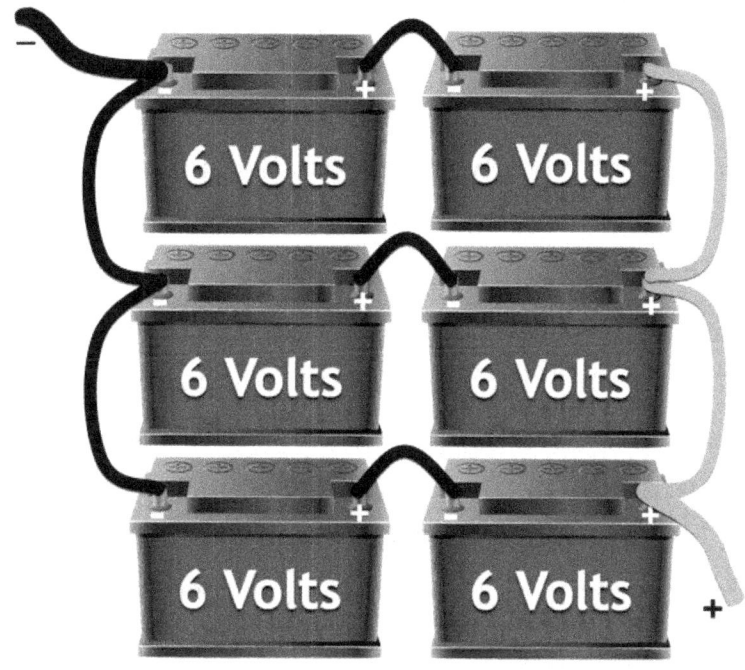

**6 six volt battery bank wired in series and parallel.**

Wiring batteries this way feels so wrong by the way. When you look at it, you might think, "Wouldn't it short something out to have a battery being connected between the positive and negative terminals and then later connect all the positive terminals together? Wouldn't that blow something up?" But no, it's just fine. That's how electricity works.

Once you've built your battery bank in this manner, you want to make sure to connect the entire battery bank to your inverter. It's important that when you do this you choose the completely opposite positive and negative terminals across the entire battery bank. In the configuration I described, that would the two extreme diagonal terminals (one positive and one negative). Doing this insures that you'll be pulling power from all of the batteries equally.

The reason this is true has to do with the nature of how electricity flows. Electricity will always take the path of least resistance. Oftentimes, you will see someone tie two wires together at each end thinking that by doubling the wire, they can double its capacity to conduct electricity. But this just isn't the case. Electricity will instantly determine which of the two wire has the least resistance and will only travel down that wire. The difference in resistance can be minuscule simply because one wire is just a little shorter than the other or it's made out of a different metal. Whatever the reason, electricity will only flow down the least resistant wire. This is also true of electricity flowing through batteries.

If you were to connect to two terminals just on one battery or one set of two batteries in series, the electricity would only flow from the battery or batteries you were connected to. The remaining batteries would then try to recharge the batteries you were using. Eventually, this imbalance would cause the most used batteries to age more quickly and overheat. In flooded lead acid batteries, the more used batteries would outgas more than the lesser used batteries and you would have to replenish them with water more often. If you see that some of your batteries use more water than others, this is an indication that you may have attached the power output on the wrong terminals.

Regardless of the number of batteries you might have in your battery bank, as long as it is set up in series and parallel to provide 12 volts, the charging mode voltages are the same as if you were charging only one 12-volt battery. You don't have to increase voltage or current to charge your battery bank just because there are more batteries.

That being said, a bank of multiple batteries will be able to accept more current while charging. Think of batteries as buckets and current as water filling the buckets. If you only have one bucket, you can only throw so much water at it. But if you have multiple buckets, you can throw a lot more water at them to fill them up. They won't fill any faster than if you charged them each up individually though. Nice try.

Following that analogy, you have two elements of electricity involved with charging; current (measured in amps) is like the water filling the buckets. The more current you have recharging the batteries, the longer they'll last. The other element is voltage. Voltage is like the pressure in the hose behind the water. If you try to fill the buckets of water with low pressure, it will take longer. Similarly, if you try to charge your batteries with a low voltage, it will take longer to fully charge the batteries.

Even worse, unlike buckets of water, batteries become more and more difficult to squeeze the last remaining charge into the batteries as they get closer to a full charge. Kind of like blowing air into a beach ball. You've really got to blow hard to get that last bit of air in it so it will bounce. In order to do this with batteries, you have to use a big voltage for a while. This is why there are different charge modes in solar charge controllers. The voltage is increased in the absorption mode to make sure the charger can squeeze in the last bit of current into the battery.

With a proper battery bank wired and connected to the inverter from opposite terminals to make sure you're using the entire battery bank, it's important to then be able to accurately monitor the voltage and battery capacity in amp hours or percentage so you can know how well your charging system is working. While some decent solar charge controllers come with a useful display to provide this information and others can have them fitted later as an extra purchase, I prefer the Tri-Metric power monitor because it is so amazingly accurate and offers a simple percentage available display that is often missing from other displays. In fact, I don't watch any other data than that simple percentage display now. I know it's super accurate and it tells me everything I need to know at a glance. I don't even have to push a button. It's just there to see from across the vehicle whenever I need to know how the batteries are doing.

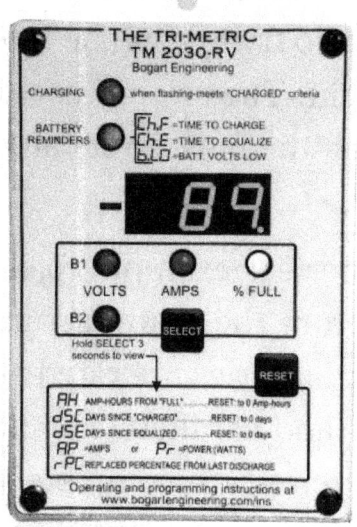

**Tri-Metric TM 2030-RV power monitor.**

Having a percentage rating like this is helpful in maintaining your batteries too because once you know that your batteries should never, ever drop below 50% and that they function best when they stay between 80% and 100%, you can more efficiently budget your electrical use. You can also quickly estimate how deep you can discharge the batteries and how much you can expect the solar charger to recharge them in a given day depending on the weather.

In my case, I know I can safely drop down to 65% and still use my crazy power-hungry instant coffee maker or microwave oven without any troubles at all. But any lower and my inverter will tap out due to low voltage. I also know that if I do drop that low in percentage, it will take two days of full sunlight to completely recharge the battery bank to 100% while I'm still using my laptop, lights and DC fans throughout the day. This often happens during the summer because I run the swamp cooler for a good part of the day and early evening to try to cool the interior of the truck so I can get some sleep at night. The swamp cooler uses about 25 amps of power at full speed on the fans, so it can drain the batteries pretty good.

In fact, this is a perfect example to illustrate how stated amp hours on a battery from the manufacturer and real-world usable amp hours differ. My batteries are rated at 220 amp hours each. By wiring them in series and parallel, I should have a total of 660 amp hours before the batteries are completely discharged. By running the swamp cooler from 10 am to 9 pm during the Summer, I'm using 11 hours at 22 amps each or 11 x 22, which equals 242 amp hours. I'm also running a laptop, another DC fan, an LED light, a coffee warmer and short but potent bursts from the coffee maker and microwave oven throughout the day. Best estimate of total amp hour use with all of that is about 450 amp hours. Then add on another eight hours with the DC fan on while I sleep and I'll wake up the next day with batteries at 65%.

However, I know just making breakfast and coffee with the microwave oven and coffee maker, I will easily drop that percentage another 2 points to 63% and if I check the voltage on the battery bank at that point, I'm starting to get dangerously close to tripping the voltage limiter on the inverter. By dangerously close, I mean if it was a socked-in cloudy day with no direct sunlight, I would have to seriously budget my electrical use or I might find myself in a low-power, lights only scenario for the night without a laptop. So, even though I should have 660 amp hours of stored power, by the time I start sneaking up on about 500 amp hours, the voltage is getting close to being too low to safely use without the inverter shutting off.

## Batteries Recap

Buy only deep cycle batteries. If the words "hybrid", "cranking", "starting", or "dual-purpose" are on the battery, move on. Words like "Golf cart" and "fork lift" are usually okay though. There are three battery choices; flooded lead acid, AGM ("no-maintenance"), and lithium. Lead acid are the cheapest, AGM more expensive and Lithium are through the roof. Lead acid batteries require constant maintenance with water replenishment and protection from outgassing and must be mounted upright. AGM batteries require no maintenance and can be mounted on their sides, but they don't have as much storage capacity as lead acid batteries. Lithium batteries require no maintenance and recharge faster than the other batteries. They're just crazy expensive. I prefer the cheapest option; flooded lead acid batteries. They're easy to find, easy to maintain, and cheap to replace. You just have to pay close attention to the water levels and avoid corrosion. It's not that tough really.

Only use 12-volt equipment. 24 volt and 48 volt systems are for residential installations and the equipment costs way more than you need to spend for a robust mobile installation.

12 volt batteries are really 12.7 volts. There are four charging stages; bulk, absorption, desulfation (or equalization), and float. Each stage uses different voltages and current levels in order to charge the batteries as quickly and safely as possible. Bulk ramps up

the voltage with constant current until it reaches 14.8 volts. Absorption maintains 14.8 volts while the current is decreased as it reaches the level of amps the batteries can handle. About once a month, the desulfation stage will kick in to make sure the batteries do not build up crystalline deposits on the plates by charging at 15.5 volts for about 3 to 4 hours. Finally, the float stage will maintain 13.8 volts at low current just to make sure the batteries are topped off.

Total amp hours of your battery bank are misleading because you will never be able to use all of the power due to voltage drop when the batteries are approaching full discharge.

When connecting to the battery bank, make sure you use the entire bank of batteries by connecting to the two furthest opposite positive and negative terminals on two separate batteries.

Use an accurate battery monitor and try to budget your electricity use to maintain the batteries between 80% and 100%. Try not to allow your batteries to drop below 50% very often or for very long. This allows sulfation to build up, which will prematurely kill the batteries.

Your inverter should never, ever shut down due to low voltage.

Phew, that was a lot. So now let's talk about solar charge controllers.

# Solar Chargers

Solar chargers come in three types; 1) worthless, 2) marginally functional, and 3) very expensive. The first two types are both PWM or Pulse Width Modulation chargers. They're not very intelligent. Essentially, they take the power from the solar panels and turn it into a set charging voltage and then use whatever current is left over to charge the batteries.

As if that wasn't bad enough, the worthless models fail to set the voltage high enough to push a useful charge while also being limited by a small amount of current they can handle. Any PWM solar charger rated at 15 amps is completely worthless unless all you care to do is run a radio on the beach. Even the 30 amp PWM chargers are too limited to be of any real use. If you can find a 60 amp PWM charger, you can at least make it work well enough to be worth your while, but without hesitation I can tell you that the only proper type of solar charger to buy is an MPPT model. You'll have to pay much more for it, but it's worth it in the end. In fact, I'll go one step further and say that a proper MPPT charger is the most important element of your entire solar power system. Whatever you cut corners on, it must not be the solar charger.

An MPPT or Maximum Power Point Tracking solar charger is far more intelligent and can manipulate the voltage and current coming from the panels in order to provide a proper charge not only with adequate voltage levels and current, but also specifically configured for your type of battery and the present charge level.

I first purchased a 60 amp PWM charger for about $80 to try to save money. For the first two years, it only barely kept my batteries charged and never fully charged them. Part of the reason why was because of a stupid temperature sensor sticking out of the top that was supposed to limit the voltage output if the weather turned hot or increase it if it got too cold (neither of those instances ever happen in San Diego). This sensor was supposed to protect the batteries. But the manufacturer set the top voltage even without the sensor to 14.2 volts at maximum sun exposure. With the sensor, it was limited to 13.8 volts at best. Those voltage levels are almost right, but because the controller could only maintain them for about an hour each day at best, it was useless.

**Typical PWM solar charger.**

Even with one of the most expensive PWM chargers and removing the unnecessary temperature sensor that was limiting power output, my power system was NEVER going to fully charge my batteries. You need at least 14.6 volts for a proper absorption charge.

Now, with an MPPT charger, that's not a problem. The MPPT charger can swap current for voltage to make sure your batteries get a proper charge. Some of the current is used to make up for the lower voltage. You lose a bit of current in the exchange, so you charge with as much current as you have until the battery is full.

It might help to think of current like a water hose. The current is like the amount of water flowing out of the hose. The voltage is like the force behind that water pushing it out of the hose.

It's also important to know that batteries have to be charged with specific voltages for specific amounts of time. There are in fact four unique battery charging stages; Bulk charging, Desulfation charging, Absorption Charging, and Float charging. I covered this in detail in the chapter entitled Battery Charge Expectations earlier.

A PWM charger is not smart enough to know the difference between these four charging stages. It simply takes power from the panels and dumps it into the batteries as best it can. An MPPT charger knows the existing charge of the batteries and is smart enough to know when to use more or less voltage and when to dump as much current into the batteries as possible. Without this smart approach, batteries would either be undercharged, causing damaging sulfation on the battery plates (which kills batteries quickly) or overcharged causing the batteries to overheat and potentially be damaged. Although, overcharging with a PWM is far less likely.

Bulk charging occurs when the sun first comes up and the charger starts to ramp up the voltage to charge the batteries with a steady flow of available current. The charger will continue until the voltage reaches the battery type's absorption voltage. For example, for flooded lead-acid batteries, that's around 14.8 volts.

At this point the charger switches to absorption charge mode. The voltage is maintained at a steady level with any excess voltage converted to current and added to the rest of the current flowing from the solar panels. This charging mode will continue until the batteries are fully charged.

Once the batteries are fully charged, the charger will ramp down to a float charge at a much lower voltage and current in order to maintain the batteries at a fully charged level as long as it can until the sun goes down.

I left the desulfation/equalization charge mode until last because it's a special charging mode that is only used about once a month. Sulfation is what kills batteries. When batteries are not fully charged, they begin to accumulate deposits of crystalline lead sulfate on the plates. If the batteries don't get a full charge for a long time, these deposits will build up on the plates reducing their effectiveness. Eventually, these deposits can completely clog up the plates and short them out. Once a battery starts to build up sulfation, the life expectancy of the battery is quickly reduced.

The Desulfation charge mode takes care of this. Once a month, the charger will wait for the batteries to reach absorption mode, but instead of keeping the voltage at a steady charging level, it will increase the voltage to the batteries maximum safe charging level (on flooded lead acid batteries, this is about 15.5 volts). It will maintain this extreme high voltage for about three or four hours and then drop back down to a normal absorption charge mode and eventually to a float charge mode as usual.

By doing this, the charger is basically "burning off" the sulfation deposits that may have built up on the plates over the past month. Actually, it's breaking up the crystalline lead sulfate into its original components of lead and sulfuric acid. In fact, that's how batteries work. Without the process of sulfation, batteries would not be able to deliver a charge. In small, controlled amounts, sulfation is a good thing.

PWM chargers don't know how to do any of this though. I learned this the hard way with my first set of batteries. They only lasted about three and half years because over time without ever receiving a full absorption charge and never receiving a desulfation charge, the deposits built up so bad that the plates shorted out, killing the batteries. (In my case, I was able to get another six months out of the batteries by trying to resuscitate them with a baking soda wash and new acid. You can read more about that in the chapter on Battery Resuscitation Tricks.

Solar charge controllers are rated by the amount of current or amps they can handle. There are 15, 30, 45, and 60 amp models in both PWM and MPPT. Most of the Harbor Freight specials have 15 and 30 amp controllers. These are basically useless. They don't manage near enough power to be useful by themselves. You could combine them to charge batteries in parallel, but that's just silly unless you're trying to really cut corners and approximate some kind of clever minimal redundant power charging solution.

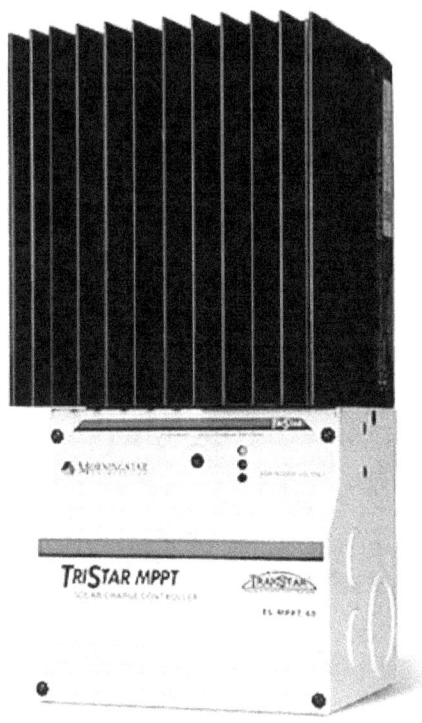

**Morningatar TriStar MPPT solar charge controller.**

Most MPPT chargers don't bother with anything less than 45 amps. This is in fact the ONLY place where you might be able to save some money on an expensive MPPT controller. If your power needs can be managed by a less expensive 45-amp charger, that's great. With my setup using a potential of 1,100 watts on the roof, I had no choice but to buy the more expensive 60-amp model.

But I didn't pay full price. Thank the Fates for eBay! Right now, a proper Morningstar TriStar 60 amp MPPT charge controller without display runs just shy of $600. I was able to find one still new in the box out of Florida on eBay with a winning bid of $450. That's about the same price for a 45-amp model new. So, I lucked out. I later bought the $99 LED display attachment, but quickly found I didn't need it because the Tri-Metric power monitor I already had was providing more than enough information. In fact, as silly as it may sound, I really only use one function from the Trick-Metric; the percentage of charge. I know it's hyper accurate and it simplifies everything down to just one very easy display. All I have to do is glance across the truck and I know instantly how well my power situation is.

It might help if I explain how I determined that I needed the 60-amp model over the 45 amp. First off, it's important to know that even though your solar panels may be rated by the manufacture at a certain wattage, that number is never accurate.

You won't truly know what kind of power to expect from your solar panels until you get them hooked up. A lot of things can affect the eventual power output of your solar panel array; different sized panels, different types of panels, wire gauges from the panels to each other and to the charger, amount of sunlight, angle of the panels, etc.

In my case, using the Tri-Metric power monitor, I was able to determine that at best, my panels were putting out about 43 amps at high noon on a clear day. That's a bit too close to 45 amps, so I opted for the 60-amp model. But this tells me a lot about my solar panel configuration too. 45 amps at 14.8 volts is only a little over 600 watts. Well, what the heck? The solar panel manufactures label indicates my panels should put out 100 watts for the smaller ones and 250 watts from the larger ones. I should have 1,100 watts total. Well, chalk it up to using wires too small, lying flat on the roof, maybe a little dirt on the panels, whatever, I'm only getting a little over half of what I expected.

But as I've said before, every piece of solar power equipment has overly ambitious ratings. Inverters are the worse. My 800-watt square wave inverter could only put out about 500 watts reliably without shutting down. The manual that came with it suggested wire gauges far too small for reliable use, so I increased the cable size to a more useful 6 gauge. That improved output to a little over 600 watts, but that's still not 800. The same is true of solar panels. To get anywhere near the stated power rating, you must use decent gauge wire, keep the panels clean of dirt or pollen, angle them directly at the sun on a clear cool day, and hold your tongue just right just to get close.

One last thing to be aware of with solar charge controllers is the load connection. Most PWM chargers have three connections with two terminals each. One is to connect to the solar panels. Another connects to the batteries. And a third one is to connect a "Load". Frankly, this load connection is worthless. You should never connect a heavy load to these terminals. That means, never connect your inverter to the solar charge controller. Always connect your inverter directly to the batteries (or through the shunt to the batteries if you're using a Tri-Metric or other power monitor). The only safe load that can be connected to these terminals is maybe a light or a fan. More expensive MPPT controllers don't even bother including a load terminal, because they know it's of no use in a professional installation.

Now, if for some reason you're still stuck on trying to save money by buying a PWM controller, know this: An MPPT controller is essentially a smart PWM controller. In simple terms, Pulse Width Modulation is how solar power is turned into battery power. But a PWM controller only knows how to use what it's given. And if it's given too much power, it just ignores it. An MPPT controller is also a PWM controller, but it's smart enough to know how much charge your batteries already have, how much power is coming from the solar panels, and (most importantly) how to convert the power from the solar panels into the proper volts and amps for each of the four charging stages. In short, an MPPT controller is a smart PWM controller. So why would you want to buy a "dumb" PWM controller?

## Mounting and Placement

In all cases with solar power systems, it is important to keep the wires as short as possible. This will heavily affect your choice of where to mount the charge controller. All too often, the location of where to mount the controller is based solely on convenience in order to read the display. Consequently, extremely long wires will be used to stretch the controller all the way inside the far end of an RV to be close to some other vehicle gauges like the water and propane levels.

One of the worst locations to mount the controller is unfortunately the most common; right above the refrigerator in an

RV. Mobile refrigerators generate heat, especially when run on propane. The heat naturally rises and surrounds the charge controller with a blanket of warm air that causes an inaccurate reading from the battery temperature sensor, which if it's part of the main enclosure is worthless anyway - a proper battery temperature sensor will be a long wire attached to the batteries. Another less obvious concern to keep in mind is to make sure you don't mount the controller in a place that will be in direct sunlight. Whatever use the temperature sensor might provide would be rendered inaccurate with the sunlight heating up the sensor.

This is of course assuming that the battery temperature sensor is integrated into the body of the charge controller, which is almost always the case with PWM chargers. This is yet another reason to consider spending the extra money for a proper MPPT charge controller. Most decent MPPT controllers will have a wired connection to a temperature sensor that is mounted near the battery bank remotely. The temperature sensor is important in areas where extreme weather can affect battery charge. Cold batteries need a higher voltage to charge properly and the charge controller needs to have accurate temperature information to determine the proper voltage. The same is true for extremely hot weather. Batteries in hot conditions must be charged more carefully to avoid damage or boiling for flooded lead acid batteries.

My own real-world experience has shown that the temperature sensors on cheap controllers, which are usually just a little dongle made out of a power type plug adapter fitted with a thermistor inside, are completely worthless. With my first 60 amp PWM controller, the manual was written in very poor English, obviously translated from another language. All it said was "You must for to insert temp sensor on casing top." It didn't indicate why or when it might be important. So, naturally I inserted it and forgot about it. The charge controller was separated from my battery bank by an insulated wall in my truck, so whatever temperature it was sensing was useless information in the first place. But that's not the worst of it.

As this was my very first attempt at building a solar power system, I didn't realize what kind of voltages I should expect to be coming from the charge controller to the batteries. I just assumed it knew better than I did what was needed. I was dead wrong - as in constantly dead batteries from undercharging.

As I read and researched more about proper charging modes and voltages, I realized that my cheap PWM charger wasn't getting anywhere close to an absorption voltage required to properly and fully charge the batteries. I was perplexed. Why would a company go to the trouble to design and build such a specific piece of equipment that doesn't actually do what it's supposed to do? That's like buying a toaster that doesn't actually toast bread.

So one day, I started poking at the control settings. There were few options to alter, none of which had any control over the actual charging mode or voltages. And then, out of shear frustration, I just pulled the temperature dongle out of the top of the controller and suddenly the display showing the voltage going to the batteries shot up from 13.8 volts to 14.6! What the heck?! Consequently, I tossed that stupid temperature dongle away.

Now, I live in San Diego, on the coast. It can get pretty hot during the Summer inland, but on the coast, it rarely gets above 80°. The Winters are pretty mild too. A really cold day at the beach in San Diego is low forties. And that's pretty rare too. So, I really don't have much need for a temperature sensor. But if you live someplace where the weather is described using terms like snow, sleet or ice or conversely, the weather report is followed by public service advisories on where you can find free access to air-conditioning at malls or libraries while the asphalt outside turns into lava, then you should definitely use a battery temperature sensor. With cheaper PWM charge controllers, that means mounting it right next to the batteries and hoping that little dongle works properly. Not your best option. With a decent MPPT controller, you can mount it in a more convenient location and then connect the wired temperature sensor remotely.

Regardless of which type of controller us use, it is important to keep the charging wires as short as possible to the batteries. The reason for this is simple; wires lose voltage the longer they get, especially with lower DC voltages over thin wires. This is why it's important to use heavy gauge wires at least 6 gauge or better as short as possible from the charge controller to the batteries.

In my truck, I mounted the MPPT controller just on the other side of an insulated wall from where I keep the batteries up front. I used 6 gauge wires and they're a little over four feet long - just long enough to fit through the insulated wall and reach the two extreme battery terminals on the battery bank. I have the battery temperature sensory connected, but with little change and lack of extremes in the weather here in San Diego, it really makes little difference in my case.

**Solar Chargers Recap**

Don't bother trying to save money on cheap PWM charge controllers. Buy a proper MPPT charge controller. I highly recommend Morningstar. (And no, I'm not a sales rep.)

Have a clear understanding of charging modes and voltages so you can properly determine if your charger is working properly.

Mount the controller as close to the batteries as possible with wires no smaller than 6-gauge.

Temperature sensors are really only important in areas with extreme temperatures. With cheap PWM controllers, the temperature sensor will likely be a dongle plugged into the top of the case. This sensor will only work well if mounted next to the batteries, away from any heat sources, and out of direct sunlight. Proper MPPT controllers will have a remote wired temperature sensor, giving you a little more leeway in where you mount the case.

Keep the wires from the charge controller to the batteries as short as possible to avoid voltage drop.

Next, Inverters.

# Power Inverters

AC (Alternating Current) power inverters are devices that convert DC (Direct Current) power from your batteries to AC power similar to that found in a common house wall outlet. Inverters come in two types; square wave (or modified sine) and pure sine. They also come in two categories; worthless and expensive. There are worthless models for each type just as there are expensive models for each. Regardless of the type of inverter, the cost is the real issue. Basically, you get what you pay for.

**3,000 watt pure sine inverter/charger.**

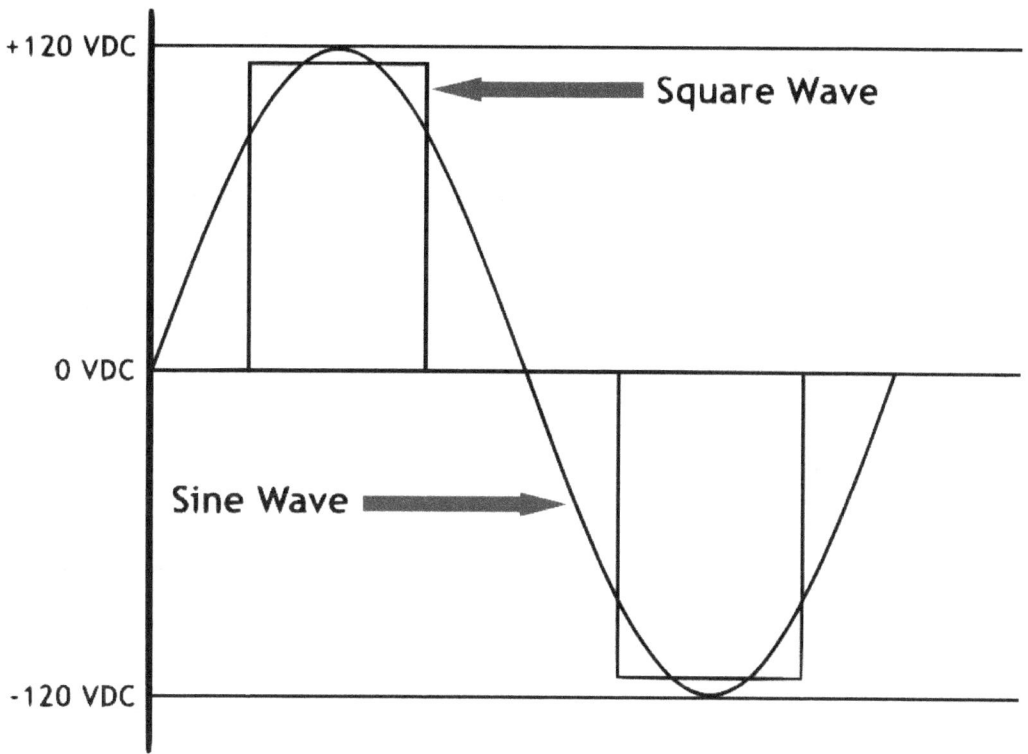

**Square (modified sine) wave vs. pure sine.**

Pure sine inverters always cost more than square wave inverters of the same power rating. This is because pure sine inverters generate electricity that is much more like the kind of clean, reliable electricity you're accustomed to in a standard house wall outlet. Square wave inverters try to save money by approximating the alternating current. It's the same voltage and they can run ALMOST the same equipment but the power isn't as smooth as with a pure sine inverter. Where this becomes important is when you want to run a microwave oven, large screen television, some power tools, or high-end electronic equipment. The square wave inverter will make these items buzz and can, over prolonged use, actually damage the equipment.

I use a 3,000-watt square wave inverter for everything except my microwave oven because it puts out a horrible sound unless I run it on my other 3,000-watt pure sine inverter. My 32-inch flat screen TV also buzzes slightly, but it's hardly noticeable. I also have a four-drive RAID array storage unit for my computer that makes a slight buzzing sound. I decided long ago to take a chance on those two items just so I could use the pure sine inverter only when I needed it for the microwave oven. My pure sine inverter also has an integrated battery charger that I can connect to a generator or larger 25-amp wall outlet if I end up having to go without sunlight for several consecutive days and have to charge the batteries back up.

After all is said and done, unless you plan to run a microwave or maybe an air-conditioner or a home computer with an expensive monitor, I would suggest going without a pure sine inverter if you're on a tight budget. If you can afford one though, by all means, go with a pure sine. The power is much cleaner and more like the kind of power you and your equipment are used to from a standard wall outlet.

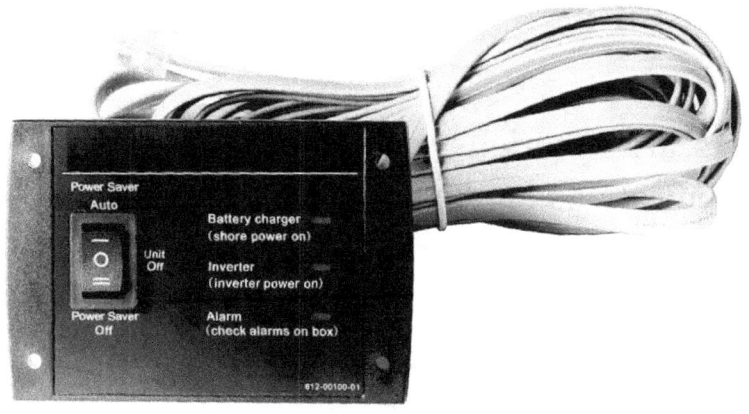

**Simple remote switch. Boring!**

**Remote outlet and switch. Way better.**

One nice feature to look for when buying an inverter is a wired remote control switch. These usually connect to the inverter with a long standard power cord and telephone like connector allowing you to mount the remote switch as much as ten or twelve feet away from the inverter. This can be really handy for shutting off the inverter (and the lights at the same time) while in bed. The remote switch might also include a power outlet and even a handy USB charging outlet for your cellphone.

## Specifications

Inverters are notorious for being far less powerful than their stated specifications on the box. This is because of two things. The first is that people don't use large enough gauge wire for the batteries to the inverter. Anything less than 1/0 gauge on a 2,000 or 3,000-watt inverter will prevent it from pulling a full amount of power quickly enough through the wires to operate efficiently. You might get away with 4 or 6 gauge for a smaller 800-watt or even a 1,500-watt system, but not for the larger powered inverters.

The second reason is that, as I've said before, every piece of equipment in a solar power system is conspiring against you. The manufactures will slap a wattage rating on the box, but that's almost always a very inflated number. With a properly wired inverter and adequate cooling and ventilation, you'll be lucky to get 80% of the stated power rating of even the most expensive inverters. That's just the way of things unfortunately. With inadequate gauge wiring and poor ventilation, it is very conceivable to get only 50%!

The big question about inverters is how much power you actually need. Once you determine your probable power usage level in watts, triple that value. Some would say to just double it, but since inverter ratings are so inflated, you would really only be getting by at best. By tripling the value, you can be assured that you not only will have as much use of the inverter as you will need, but you'll also have some leeway for sudden power spikes from things like coffee makers or power tools as well, as some room for growth should you decide to add more things like extra lighting or fans.

# Floating Ground

Another issue to be aware of with inverters is floating ground. This is a problem that can plague even some of the more expensive inverters. The problem is that a mobile vehicle has to make do with using the frame of the vehicle as a reference power ground. You don't have a spike driven into the actual ground like with a house. This ground is different from the negative "ground" wire of the batteries. And therein lies the problem.

Some inverters handle floating ground better than others. And it only has slightly to do with the cost. It also has nothing to do with whether the inverter is a square wave or pure sine. It's all a matter of whether the manufacturer has designed the inverter well enough to handle grounding properly. It's as simple as that.

If you have an inverter that doesn't handle floating ground properly, it's not particularly dangerous, but it can cause some equipment to function erratically or not at all. Two perfect examples of this are touch pads on laptops and some wireless network hubs. The touch pads require a reference ground to compare to the minuscule current that flows through your body as you touch the touch pad with your fingers. Without a proper reference ground, it can't accurately measure the changes in current and the mouse pointer will either not move or it will move in a jerking motion or shoot across the screen uncontrollably.

With wireless network hubs, the signal is "dampened" by a lack of reference ground and wireless connection range can be severely limited. It can also have difficulty connecting over network cables between devices because each device's power source will be trying to work with the grounding issue as best it can and there are bound to be some differences between each that cause communications to be disrupted. I've even had a network dongle overheat because the laptop was on one inverter and the other computer was on another inverter and they couldn't agree on a common ground.

Unfortunately, if your specific inverter doesn't handle floating ground properly, there is no simple solution to correct this problem. You may be able to reduce the effects of the problem by grounding the outside of the case to the frame of the vehicle. If you do though, be sure to keep the connection to the frame far from the connections to the vehicle battery or your solar battery bank. They are not the same ground.

I have a relatively inexpensive (~$300) 3,000-watt square wave inverter that I use for everything but my microwave oven and it manages floating ground perfectly. But I also have a moderately expensive (~$1,000) pure sine inverter that is great for running my microwave oven and doubles as my AC battery charger if I ever need it, but it's horrible at managing the floating ground. I can't run my laptops or network off the pure sine inverter at all - an inverter I paid over three times as much for! Go figure. In fact, of all of the inverters I have owned; 100-watt square wave, 800-watt square wave, 1,500 watt pure sine, 3,000 watt square wave, and a 3,000 watt pure sine, only the 3,000 watt square wave model had a proper floating ground.

The lesson to be learned from this is to buy the highest quality inverter you can afford of the type you need (pure or square wave) and test it out first. If you can't get it to manage floating ground to your satisfaction, send it back. It's hit or miss. Sadly, that's just the way things are.

## Inverters Recap

Inverters come in two types; square (or modified sine) wave and pure sine. The pure sine are more expensive because they more accurately simulate the power you would get form a standard wall outlet. You only need a pure sine inverter if you're running expensive equipment that needs it. The easy way to tell is if your devices make a buzzing sound while connected to a square wave inverter, and it's expensive stuff, you might want to consider a pure sine wave inverter. Microwave ovens, big TVs and some electronic equipment usually need pure sine. Damage can occur if you run equipment on a square wave that needs pure sine.

Try to buy an inverter with a wired remote switch and outlet. They're really convenient.

Always triple the wattage rating needed when buying an inverter. The manufacturers lie about the true power capabilities and you need room for power spikes and additional items added over time that doubling the rating might not cover.

If you use a laptop with a touchpad or a wireless network hub, you may have to test your inverter for proper floating ground or things might not work as expected.

Now let's talk about how to wire all of this stuff together.

# Wiring

Wiring is one of the most hotly contested and debated aspects of solar power systems. It's also one of the most confusing, which adds to the debate.

The argument stems from a fact about electricity and DC current. Low DC voltages will lose volts over long lengths of wire and/or thin wire. The problem is, installers can't agree on what "long" and "thin" mean. Part of the problem is that some of them are talking about 12 volt systems while other are concerned with 48 volt systems. And both are interested in cutting corners and saving money whenever possible. Thinner wire is less expensive and much more flexible, making it easier to route through a vehicle than the heavier 4 and 6 gauge wires that are actually needed. And while there is a slight difference between 12 and 48 volt systems, it's not enough to justify using such tiny wires.

The longer a wire is, the more power that can get lost along the way. Wire is a conductor, but it's not a perfect conductor. So, the longer the wire, the more resistance the power has to run through to get to the other end. Low voltage DC systems (that includes 48 volt systems!), are very susceptible to power drain with long lengths of wire. So much so that a 14.8 volt float charge from the solar charge controller could drop to 14.2 volts by the time it reaches the battery making it "don't even bother" charge. Sure, it'll charge the batteries, but not near as well as it should just because somebody chose to cut corners and try to save a few bucks on wire.

The bigger concern with using thinner wire is current flow (amps).

If you run high current through a thin 10-gauge wire, the wire will heat up significantly and can easily melt the insulation off and eventually short out on the frame of the vehicle or ignite the insulation itself and catch fire.

Compounding the problem is the fact that wire gauges are numbered weird. The American Wire Gauge (AWG) scale assigns larger numbers to thinner wires and smaller numbers down to zero for larger ones until you get to 0 and then it gets even weirder with 1/0, 2/0, etc. for the really big stuff. The larger wires often refer to the zero as "aught". So, 1/0 is single aught, 2/0 is double aught, etc. Nuts!

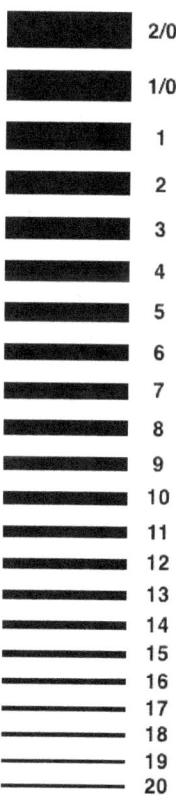

**Wire gauges (not to scale).**

What's worse, equipment manufactures really muddy waters with mixed messages and inaccurate statements concerning wire gauge requirements. I had an 800-watt inverter that insisted that only 4 gauge or larger wire be used but then came equipped in the same box with 8 gauge wires! I also had a 100-watt inverter that had a cigarette lighter plug using 14-gauge wire but with a 5 amp fuse (5 amps x 12.5 volts = 62.5 watts). The fuse would blow if I actually tried to use 100 watts. As if that wasn't bad enough, the cigarette lighter plug on the car was wired with 18-gauge wire which is far too small to support any kind of useful current in the first place.

On the larger scale of things, I have a 3,000-watt inverter with instructions that insisted only 0/0-gauge wire be used. Okaaaay, I know there's 1 and there's 1/0. What is 0/0? Did they mean "oo"? If so, that's actually 2/0. And that stuff is a third of an inch thick! It's also extremely expensive. It also didn't help that the connectors on the inverter itself were too small to connect to such a large cable.

I had a similar problem with a 1,500-watt pure sine inverter that came with tiny wires that could never support that much power. It also didn't come with any instructions whatsoever.

Another potential gotcha (and much contested issue) with wiring you might not even expect is uneven lengths. You should keep the positive and negative lengths for the same connection as equal in length as possible. This is often pretty easy to do, but sometimes, especially when wiring to the opposite ends of a large battery bank, you might be tempted to use a shorter length on one end because the equipment connected to it is closer on that end. Try not to do this. Also, never, ever coil up excess wire - not because coiling is a problem, but because you shouldn't have any excess wire! So how are you supposed to do both? Just do the best you can.

The explanation for this issue is a bit long-winded, but bear with me. Electricity will always flow along the path of least resistance. Two wires of different length will have two different levels of resistance. The shorter wire will have less resistance than the longer one. Okay, nothing surprising there. Where it gets weird is when you realize that the flow of electricity is not really a flow as much as it is a kind of bucket brigade of electrical atoms passing a charge from one atom to the next in a long line down a wire. Shorter wires have fewer atoms. This equates to less resistance, so as we follow the analogy, the shorter wire bucket carriers work just a little bit faster than the ones on the longer wire. When this happens, the shorter wire atoms will be "pulling" for more power than the longer wire can provide. And since electricity always travels through the path of least resistance, the "faster" atoms will pull a charge from anywhere they can find it. They will kind of "grab" at other nearby atoms. This occurs most where other resistance issues also exist such at the terminals on the battery. The terminals may be made of a different metal than the loop on the wires which may also be made of a different metal than the wire itself which may also be made of a different metal than the nut holding it all down on the terminal itself. That, along with dirt or oil between the connections on the terminal, moisture, or even a slightly loose connection can all create minute amounts of resistance, just enough for electricity to notice.

Why is this important? Because if you have an unbalanced wire setup with different resistance levels, the constant flow of electricity can eventually eat away at the least resistant component. This can result in a process called electrolysis. It only happens in direct current (DC) circuits as opposed to alternating current (AC) circuits because electricity only flows in one direction. As it does so it may pick up stray atoms from the least resistant metal and carry it away. Yeah! Seriously. In my case, the nut on the negative terminal was the weakest length. Over four years with a much longer positive connection wire to the battery bank (and some built up corrosion from outgassed hydrogen binding to the terminals), the nut was gradually eaten away until it no longer fit the socket and had to be removed with pliers and replaced. Crazy, right? None of the other terminal nuts had this problem.

Mythbusters on the Discovery Channel actually did a show using this exact DC power electrolysis process to eat away at prison bars!

So, keep your wires short and as close to the same length for both positive and negative in the same circuit. It seems like a ridiculously small thing to worry about, but it can be a real issue.

One final concern about wires to keep in mind is the type of strand. Automotive sound system installers like to use fine wire strand because the wires are much more pliable than the heavier thick strand wires you might find at an automotive store. As far as electricity concerned, as long as they are the same gauge in size, there is no difference in efficiency, performance or safety between the two. Use whichever you prefer. The thin strand is easier to work with, but it usually costs a little bit more. It also comes in colors other than red and black, so if you want to get cute with your installation, there's that option.

One type of wire you should NEVER use for mobile power systems is solid strand residential wire. You'll find this kind of wire at home electrical shops or hardware stores. It is not a safe option for mobile systems because any constant movement from a mobile vehicle along the wire can eventually cause it to break or fracture inside the insulation. You would never see the break because it would be hidden by the insulation and it would be a pain in the neck to find it in order to replace it. But even more important, if the wire were to fracture just enough to maintain a connection, the amount of resistance generated by a wire fracture could cause the wire to overheat and catch fire. So don't do it. Solid wire is a bad idea.

So with all of this information, how is anyone supposed to be able to figure out which size wire to use for what? Heavy gauge wire is expensive. A couple of lengths of 1/0-gauge wire can cost as much as a cheap inverter! It will all make more sense in the next section, so read on. (Okay, it might not make more sense, but it will be based on real-world experience and not some badly translated manual).

## Which Gauge for What?

The solar panel manufacturers have standardized the connectors and wire size on the panels themselves to 10 gauge, so that's been decided for you. But that can be a bit deceptive. It leads installers to think that 10-gauge wire is acceptable for the entire power system. It's isn't. Not by a long shot. It's certainly not acceptable if you have a large solar panel array capable of generating high current. If you have more than 500 watts on the roof, you should connect the wires to a buss connected to at least a 6-gauge wire when dealing with several panels connected in parallel. Otherwise, your wires WILL overheat at the charge controller which can lead to a fire hazard. (Yep, done that. Melted the crap out of my cheap MPPT controller. Movin' on...)

That leaves the wires from the solar charger to the batteries, the batteries themselves, and finally the connection from the batteries to the inverter. Each connection delivers a different amount of current and therefor requires different size wires. The wires from the solar charge controller to the batteries should be at least 6 gauge. Wire that size is fairly easy to find at hardware and automotive stores. If you have a really large solar panel array (over 1,000 watts), You should increase the wire size to at least 4 gauge both from the array to the charge controller and from the charge controller to the batteries. Unfortunately, some charge controllers use terminals that are too small for 4-gauge wire in spite of the controller being rated for the high amperage. This is a bad sign that the charge controller manufacture may be cutting corners or not even know any better.

Next are the wires used to interconnect the batteries. If you have just one battery, then of course this is a non-issue, but if you have multiple batteries in a bank, it's important that you not only use the proper size wire, but that you keep it uniform in size and length between all of the batteries. If you don't, your batteries will not be charged evenly. This can cause a lot of problems, particularly with overheating, corrosion and sulfation. Automotive stores sell short (10 to 12 inch), pre-manufactured 4 or 6 gauge cables with loop ends that are perfect for this. I prefer the larger 6 gauge cables myself.

They're a lot easier to find and cost a lot less while still doing the job just fine.

The most important length of wire you'll have to deal with are the wires from the batteries to the inverter. These wires need to be VERY large. I recommend no smaller than 1/0 gauge. An inverter larger than 2,000 watts should ideally use 2/0 gauge, but that stuff is expensive and heavy. The wires should be equal length on both the positive and negative side and of course be as short as possible to allow for maximum power delivery while also allowing you to mount the inverter in a cool, dry place away from any battery fumes if possible. An ideal arrangement would be with a wall separating the batteries from the inverter and a ventilated enclosure for the batteries. This is very important because of outgassing. If you decide to save money and purchase flooded lead-acid batteries (as I recommend), it is expected that the batteries will heat up during charge and corrosive gas will escape. It's tough enough keeping the battery terminals and wires from corroding without having to worry about your expensive inverter getting eaten up inside as well.

You will also want to create a dedicated DC connection for your 12-volt power needs. Cheaper solar chargers have a feed-through DC voltage connection for this purpose usually labeled "Load". While this is convenient, some of the more expensive MPPT charge controllers don't bother with separate load terminals because the manufacturers expect you to get your DC power from the batteries directly. Wiring for low-power DC loads should be no smaller than 10 gauge. This includes DC fans, refrigerator pilot light controllers, etc. Under no circumstances should you EVER connect an inverter or other high power electrical device to the Load terminals. These terminals are for low power consumption devices only.

It seems counter-intuitive to have to consider four different wire gauges when designing and building a solar power system, which explains why so many (supposedly reputable) installation companies get it wrong somewhere down the line. Remember that decked out RV with the $80,000 solar power system I mentioned back in the chapter on Battery Charge Expectations? Well, here's the rest of the story because it's all about wiring.

As I mentioned, I saw this huge RV in Arizona decked out with a massive solar power system with all the trimmings that was basically a residential 48 volt, grid-tie system built on top of an RV.

It was squeezed into a gas station in Arizona and I just had to introduce myself. We talked about twenty minutes as Chuck, the proud owner, gave me the grand tour. And the saddest comment he made was that he had to run the generator every night because the batteries were going dead. I was flabbergasted until I saw why. They had completely wired the system with 10-gauge wire! I couldn't believe my eyes. All of this expensive equipment and it was going to waste. They had actually convinced this poor guy that solar power just "wasn't ready for prime time" and he would never be able to live solely on solar power no matter how much he paid. You can imagine the look on his face when I explained to him just how wrong they were.

I ended up giving him a tour of my lowly truck. I easily had half the number of solar panels on my roof, and I've only paid a total of about $3,600 for everything (including a $900 generator I no longer need). When I showed him my coffee maker and microwave oven, he was amazed and somewhat incredulous. It was cold outside, so I offered to make him a cup of coffee. He agreed, and within a short couple of minutes, we were both enjoying a hot cup of premium dark roast. He confided in me that he couldn't even run his drip coffee maker without the lights dimming. I just shook my head. I wanted to cry for the poor guy. I showed him in my truck the kind and size of wiring he should be using and then we went back to his RV and I showed him point blank each wire that should be replaced. The huge solar panel array was choked by the standard 10 gauge wires and the power just wasn't getting to the batteries. When I touch the connection to the solar charge controller, the wires were hot. The same was true from the batteries to the inverters; 10-gauge wire.

I gave him my card, but I never heard from him again. I hope he's okay. Chuck, if by chance you're reading this (or someone who knows Chuck), please give me a ring! I'd love to hear how you're doing.

I know what happened here. The logic of the people who installed his system was that higher voltage system don't require wiring as heavy as lower 12 volt systems because there is less voltage drop when working with higher voltages. While that is true in theory, the logic falls apart when considering current (amperage). Never mind they were also using extremely long runs of this light-weight wire across the length of the RV. Long, thin wires will choke the life out of a power system, especially between the batteries and the inverter.

By using heavy gauge wires as short as possible for the solar charging side and inter-battery connections and even heavier gauge wires for the inverter, your system will be able to move both voltage and current efficiently without any bottlenecks which cause power drain and overheated wires that can lead to fire. There aren't a lot of places where you can cut corner when building a solar power system. Wiring is one of them.

## Fuses

How important are fuses? That depends. How do you feel about a vehicle fire that destroys everything you own, including your nomadic home, all with one quick spark? Fuses are not just for the protection of your equipment. They're to protect your life as well.

I can't stress this enough. Fuses are NOT an option, ESPECIALLY on mobile solar power systems. The reason for this is simple; motion. Vehicles are constantly in motion. Batteries shift, wires bend, solar panels move. All of this motion can lead to worn wires, loose connections, or disconnections that can lead to electrical shorts either with other equipment or the frame of the vehicle.

But it's not just motion that can cause serious problems. Rain can leak inside the vehicle and cause a short. You might at one time wash the vehicle and accidentally get something wet.

Regardless of how it happens, you need to make sure your equipment and your body are both protected. Fuses are an inexpensive solution.

You will also need more than one fuse. And each fuse will undoubtedly be rated for different current levels depending on how much power will be flowing through the circuit.

There are already some built-in safety features in the solar charger to prevent reverse polarity and other stupid mistakes, but you still need a fuse. In fact, every connection in your system between every component should have a fuse with the appropriate amperage rating. But how do you know what amperage rating to use?

First, let's list the connections we're concerned with:

1) Solar panels to charge controller.

2) Charge Controller to batteries.

3) Batteries to inverter.

4) Direct DC connection to batteries.

The purpose of a fuse to is to allow a certain level of current to flow until it reaches its limit and then break the connection so that no electricity can flow if the current gets too high. This makes the choice of fuse rating to and from the solar charger a no-brainer. Your solar charger will be rated for a specific amperage. They can be anywhere from 10 amps to 60 amps, but a robust power systems with a proper MPPT controller will be at least 45 to 60 amps. Whatever your solar charge controller is rated at is the same rating your fuse should be between the connections to and from the charge controller. Thankfully, the solar panel manufacturers take care of the panel side, so all you have to worry about is the connection from the charge controller to the batteries.

The fuse for the third connection between the batteries and the inverter is a little trickier. You have to do a little bit of math to find out how many amps and volts make up the wattage rating of your inverter. Watts = Amps x Volts.

The simple math would suggest that the Volts part of the equation be equal to 12. But you now know that 12 volt batteries are really 12.7 volt batteries. So, smart money is on the person who uses the right number in their calculation.

But there's another gotcha waiting in the wings with the inverter. Most high-end inverters will deliver near the stated wattage rating, but they can also deliver a "peak" power level for a brief moment for certain appliances that can pull a much higher current when they start up. This is often the case with motorized tools or compressors that can cause a spike in current for a brief moment. The peak current rating is often twice the normal stated wattage rating of the inverter. So, a 1,000 watt inverter would have a 2,000 watt peak. If you only "do the math" for the stated current and install a fuse for that level of power but then switch on a device that causes a quick power spike, you're going to trip the fuse.

So, once you find the "peak" wattage of your inverter and you know to use the proper voltage of 12.7 instead of 12, you're ready to do the math. For a 1,000-watt inverter, the peak will be 2,000 watts. Divide 2,000 by 12.7 and you get roughly 157.5 amps. If you had divided by 12 instead of 12.7, it would be roughly 167 amps. Either way, you need at least a 175-amp fuse. (However, in my experience, it's often easier to find 200 amp fuses.)

The forth connection is a bit more arbitrary but just as important as the others in regards to avoiding damage from shorts. You can pull a significant amount of current directly from the batteries for DC-powered equipment such as fans, lights, refrigerator control systems, coffee cup warmers, etc. To simplify things, you might consider using the same fuse you bought for the solar charge connections. This would make it easier to keep a ready backup supply of fuses of the same type while avoiding any confusion as to which fuse is for what. The inverter fuse will undoubtedly be much larger and may even look completely different. So, no confusion there.

Each fuse should be placed in a fuse holder as well so you can more easily see and replace the fuse if necessary.

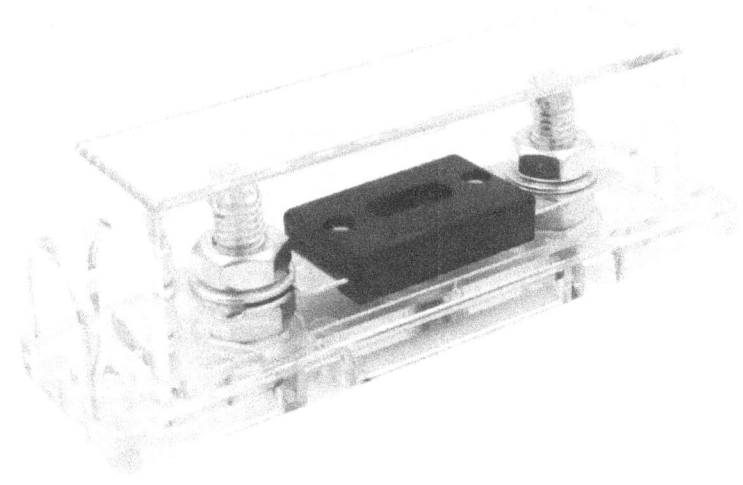

**Fuse in fuse holder.**

If you really know your way around electricity and electrical equipment, you might consider using breaker switches instead of fuses. They're more expensive, but they don't have to be replaced if they're tripped by an excessive current load. You just locate the problem that caused the breaker to trip, whether it be a short or a spike from an appliance, correct the problem, and then reset the switch. If a particular electronic device causes the breaker to switch off too often, you most likely need to recalculate the rating limit for the breaker switch in order to accommodate the device you're using. Also, be sure it's a DC power breaker and not a household AC breaker. The current rating for an AC breaker switch will not be accurate when used with DC power.

Whichever solution you choose, don't get too hung up on the precise amp rating needed. It's better to err on the higher side to avoid unnecessary fuse replacements or reset breakers. The reason for this is simple; you're not really concerned about power usage as much as power shorts. You want the power to flow as long as it behaves, but if it's racing through a short to the frame of the vehicle, you want the circuit to be severed by a fuse or breaker immediately.

**DC breaker switch.**

Whether you use fuses or breaker switches, the amount of current they can handle can handle in any of these connections is really kind of a moot issue. You're not installing a fuse to handle situations where one component is pulling too much power and might cause damage. The charge controller and inverter have built-in safety features for that sort of thing. What you're concerned about are outright shorts where electricity from one of these connections ends up going in the wrong place because of a break or exposed wiring. The choice of amperage rating for each fuse or breaker is just to make sure that the devices can function within their stated peak ratings without tripping and breaking the circuit.

## Wiring Recap

48 volts is not high enough voltage to justify thinner wire. 10-gauge wire is too thin, period.

There is no difference between fine thread wire and heavier thread, just never use solid residential wire and make sure to use the proper sized gauge wire for what you're connecting.

All wires should be as short as possible to avoid voltage drop. Both the positive and negative wires in a circuit should be equal in length to avoid electrolysis from eating the most resistant component.

The wires from the solar panels will be 10-gauge standard from the factory. If you have a large array of panels, you should consolidate the parallel wires to heavier gauge to the charge controller (at least 6 gauge - 4 gauge for larger arrays) using a wiring buss.

The wires from the charge controller to the batteries should also be no smaller than 6 gauge.

Wires connected to a "Load" terminal in a charge controller should never be larger than 10 gauge because you should never run heavy power consumption devices from the load terminals - only from the batteries directly.

Wires from the batteries to the inverters should be much larger, preferably no smaller than 1/0 (single-aught) to allow for high current flow to the inverter. Direct connections to the batteries for DC-powered devices can use 10-gauge wire, but only as long as the total current use is below 10 amps (that's not a lot).

If you use flooded lead acid batteries, it is imperative that you do not mount the solar charger or inverter in the same space. The outgassed hydrogen from the batteries will bind with the metallic the electronics and damage the equipment. Install fuses between each and every piece of equipment, including the batteries. Fuses are not an option. You can use breaker switches instead if you can afford the extra cost and know what you're doing.

Okay, we've got everything connected. Now we just need to see how well everything is working. That requires a power monitor. Next chapter.

# Power Monitors

There are various meters and indicators available to help keep you informed of the condition of your power system - more specifically, your batteries. As with all things associated with solar power systems, they range in price and value from cheap and worthless to expensive and over-the-top.

The worst indicators are the ones that are usually installed as standard equipment in RVs. These are either simple light bar meters or analog gauges.

**Simple battery light bar meter.**

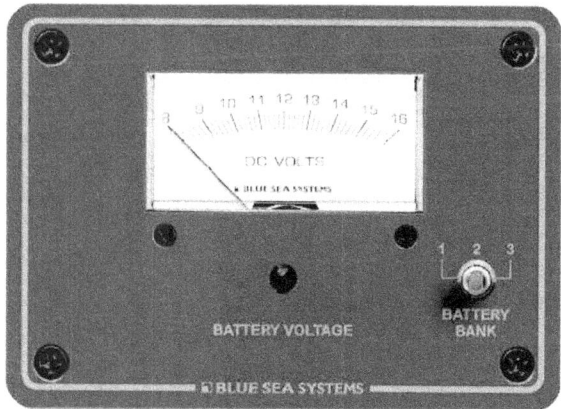

**Basic analog voltage gauge.**

In spite of these devices costing upwards of $90, both of them are far too vague in the information they provide. If nothing is drawing power from the batteries, checking these gauges might at best give you an idea of whether the batteries are dead or not, but that's about it.

Most PWM charge controllers come with more useful displays that provide information about the battery bank voltage, solar power coming in, power going out to load, etc. But the gotcha here is there is no good reason to buy a PWM controller in the first place. As I've mentioned before, the only decent solar charge controllers are MPPT. And it's very telling that some of the best MPPT chargers don't even provide a display at all. For example, you have to pay extra for a TriStar digital display from Morningstar. They offer both a front piece replacement model and a remote wired model. They provide all the information you might need, but I prefer an even more detailed power monitor called Tri-Metric.

The Tri-Metric power meter is a remote wired box you can put anywhere. It comes with a very long cable. The nice thing about the Tri-Metric is that it is designed by engineers who are serious eggheads about solar power.

**Tri-Metric ground shunt.**

The Tri-Metric tracks everything about your power system using a special shunt that attaches to the negative pole of the battery bank which allows it see every bit of power that goes in or out of the batteries. This allows it to track voltage, amps, amp hours used, amp hours remaining until full charge, and a lot more. But my favorite function that I rely on most is a simple percentage of power left in the batteries.

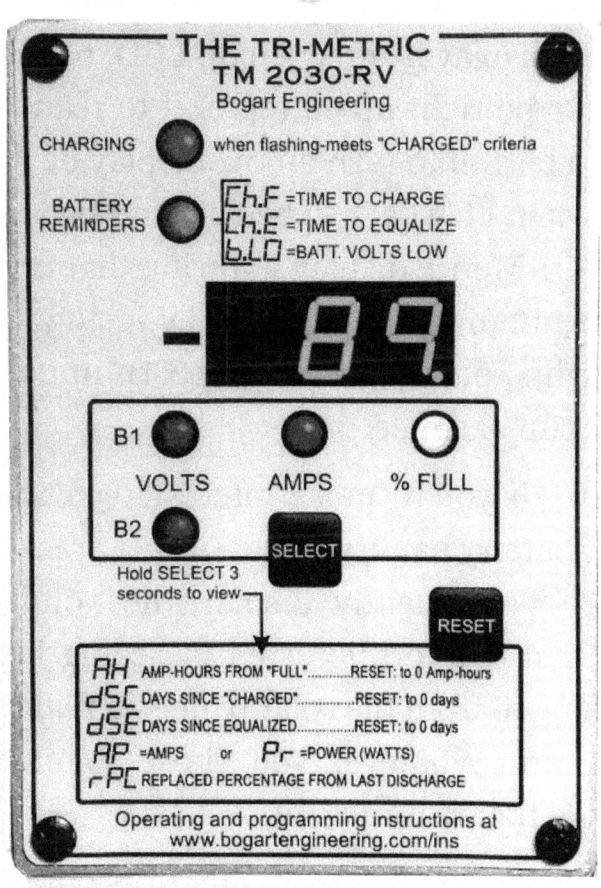

**Tri-Metric TM 2030-RV power monitor.**

It's a bit ironic that I would spend the extra money for a Tri-Metric monitor only to use one of the simplest functions it provides more than any other, but the simple fact is it's extremely accurate and easy to use. All I have to do is glance up at the monitor and I know right away, by looking at the red LED display, the percentage of charge the battery bank is currently at. I don't even have to get up out of my chair or push a button. The display is large and bright and tells me everything I really need to know - Are the batteries good? {Glance} Yep, 89%. Good to go.

With the cheap PWM controllers and even the expensive MPPT controllers, the display is often small, LCD, backlit (if you're lucky) and you have to be right up on top of it to read it. You might even have to press a button to turn on the display or the backlight. I rest easy at night when I lay back in bed around two in the morning and glance at the Tri-Metric display knowing that my batteries are still at 80% after a full night of watching TV, using my laptop, running the fan and lights, popping some popcorn in the microwave oven, and making a cup of coffee.

What's more, if you're more interested in watching the actual voltage on the battery bank, you can simple press a button to change the display to permanently display the voltage instead of the percentage. Or if you're interested in seeing how many amps or watts are coming in from the solar panels in real time, you can press the button again. Whatever you chose to monitor will remain displayed. I like that feature a lot.

I would mention other power monitors, but honestly, it's Tri-Metric or nothing. They're just that good.

## Ghost Loads

Another very useful function of the Tri-Metric monitor is the ability to monitor amperage use in real-time. This allows you to very easily track down ghost loads.

Ghost loads are power drains from equipment you didn't know were using power. This happens more often than you might think. Even though something is plugged in, if it's turned off, it shouldn't be using any power. But that is not the case. In fact, MOST electrical equipment has a ghost load when plugged in. Some exceptions are lights, fans, drip coffee makers and coffee cup warmers. Those items usually do not use any power at all when turned off. But TVs, microwave ovens, coffee makers with a clock or timer, cell phone chargers and even bread making machines and swamp coolers use a small amount of power when turned off if still plugged in. And of course, the power inverter itself will have a ghost load even when it's not running anything else plugged into it.

Ghost loads are usually very small individually (about half an amp or less), but they can quickly add up. If you have a TV, microwave oven and a cell phone charger plugged into an inverter, that could add up to a continuous power use of at least 5 amps. That's over 40 watts! Just having those few things plugged in and doing nothing is like leaving a 40-watt porch light on 24 hours a day.

So, it's important to locate these ghost loads. And just like with regular ghosts, it's a lot easier to find them at night. When the solar panels aren't charging the batteries, there won't be any amperage coming in. So, your power monitor will only be showing amperage being used instead of a net amperage between what is coming in and what is going out. By watching the amperage display and moving around the vehicle unplugging each device, you can quickly see how much power each item is using. The Tri-Metric makes this so easy because the display is so easy to read and stays on. Once you know which items have ghost loads, you can prevent unnecessary power usage by simply unplugging them or putting them on a power strip with an on/off button. Simple power strips don't have ghost loads.

In my vehicle, I have my cellphone and iPad chargers, my LED lamps and my laptop plugged in and working all day. But my TV, microwave oven, coffee maker, swamp cooler, toaster oven, bread making machine and ice maker are all unplugged or on power strips that are turned off until I need them. And at night, I turn the inverter off completely insuring there's no power use at all. Of course, I can do this at night because I have DC fans and a clock and my refrigerator runs on propane. The reason for that is even if everything is turned off or unplugged, the inverter will be using power to do nothing but run its own cooling fans. That's only about half an amp or a little over 6 watts, but over the course of eight hours while I sleep, that adds up to nearly 50 watt/hours. Leaving an inverter running just to hear its own fans for eight hours is like running a small porch light for an hour. And why would you want to listen to your inverter fans whirring all night on a warm summer night? Turn it off at night. Use DC lights to see where you're going.

## Power Needs Assessment

Having an accurate power monitor is also useful in determining your real-life power needs, especially when you buy a new appliance or electronic device.

Once you have your system put together and can monitor the power used by each item, you can quickly get an idea of just how much power you'll need to have in reserve in the batteries in order to maintain power at levels you need to get through the day and night. If all you need is a small fan, a cell phone charger and a light, you might find that your power usage is low enough to only need a couple of 6-volt golf cart batteries, and a small 800-watt inverter. But if you decide later on that you'd like to put in a TV or toaster, you might find yourself out of power sooner than you expected.

Using a power monitor, you'll be able to see exactly how much power you're using for how long and budget accordingly or expand your capacity with more batteries, more solar panels or a more powerful inverter as needed.

I started out with two 6-volt golf cart deep cycle batteries, two large 250 watt solar panels, a PWM solar charger and a tiny 800-watt square wave power inverter. This was fine for a couple of LED lamps, my cell phone and iPad chargers, a fan and occasional TV - very occasional (as in special occasions, like the Oscars or a 49ers game). There was no way I would ever run my coffee maker, microwave oven or toaster oven. It was bad enough that my cheap PWM charge controller wasn't even giving me half of the power from the solar panels, but the inverter and batteries just didn't have enough juice to get through a second cloudy day.

It took me five years to find a decent balance with all of my equipment so I could run whatever I wanted when I wanted, day or night in spite of cloudy days. I can now use my microwave oven at full power for as long as I need to. My coffee maker is an instant-hot K-Cup from Keurig. That thing uses a massive amount of power. In fact, as far as the batteries are concerned, it's like hooking up jumper cables and just tossing them into a bucket of water.

My point is, I would never have been able to do that without an accurate power monitor to help me determine my real-world power needs. I had no idea that my coffee maker was pulling over 110 amps just to make me a quick cup of coffee. That's over 1,300 watts. It never occurred to me that my 20-inch AC three-speed fan used nearly three times as much power as my 15-inch DC fan. I thought my laptop only used about 40 amps. In fact, depending on what I'm doing (watching videos, editing videos, etc.) It can use as much as 100 amps. That's a big difference.

To achieve this balance, I had to expand my battery bank to six 6-volt golf cart deep cycle batteries, two 3,000 watt inverters (one pure sine, the other square wave), a proper MPPT solar charge controller and six more smaller solar panels adding up to a potential total of 1,100 watts on the roof. In addition to all of that, I had to install a high-powered capacitor between the battery bank and the inverters to allow me to use the Keurig coffee maker because it used so much power so quickly, the batteries alone simply couldn't provide enough power quick enough, even over 1/0-gauge cable! You can read more about that in the chapter on capacitors. Just know right now that it is a hotly debated subject. Some consider the use of a capacitor as nonsense. All I know is without it, I couldn't make coffee. With it, I can. That solved the debate for me pretty quick. You be the judge. There's more to that story, but we'll get to that later.

## Power Monitoring Recap

Get a Tri-Metric power monitor. Look for ghost loads at night. Turn your inverter off when you go to bed. Use your Tri-Metric to know exactly how much power you're using so you can assess how much reserve power in your batteries you need. Expand your power system by adding more batteries, solar panels and a more powerful inverter as needed. And you might need a high-power capacitor if you're going to use some devices that are real power hogs.

Okay, so just how much power do you need anyway? Let's find out.

# Determining Power Requirements

This should be easier than it is, but since you can't trust manufacturer information regarding power usage of appliances or power capacity from inverters, you're left with having to make an educated guess. And getting the information you need is tricky because you either have to just estimate and hope or build an entire solar power system with a decent power monitor and measure each and every device you own. In the end, you actually end up doing both. You try to guess, and then when you build it, you either learn the hard way that you overestimated (and probably spent more money than you needed to) or you underestimated and you're reading this book by candlelight.

I've said this before, but it's worth repeating; you should triple your power usage and capacity estimates, not double them. Every book I read on the subject of solar power systems that even remotely discussed mobile installations suggested that doubling your estimated power needs would be sufficient. And while some of them discussed the fact that manufacturer labels can be misleading for power usage, none of them mentioned the same fact about power capacity of inverters. And that's really important.

Inverters NEVER run at the capacity indicated. Even with proper, heavy gauge wiring, short length wires and new, fully charged batteries AND the solar panels currently charging, the best you might get from ANY inverter is about 80% of stated capacity. So, an 800-watt inverter will actually only provide a little over 600 watts. My $1,000 3,000-watt pure sine inverter/charger starts to choke at about 2,500 watts. It's annoying, but that's just the facts.

So, if you estimate your power needs by reading inflated power usage ratings on manufacturer labels and then trust the stated power capacity of your inverter, and if you're very lucky and decided to double your estimate and bought twice as much inverter power, you might just break even. But that is no way to design and build a power system you'll have to depend on for daily use.

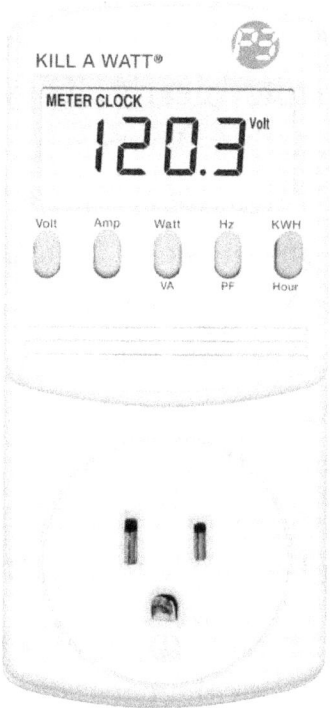

**Kill A Watt Electricity Usage Monitor.**

One useful way to get accurate information on power usage for each electrical device you'll use is to buy a Kill A Watt monitor. They cost about $20. You plug them in between a wall outlet and the device you're monitoring. Of course, you have to have a wall outlet to do this. Alternatively, if you have already built a functioning power system with a proper power monitor like the Tri-Metric, you can easily just plug in the device and watch the power meter display in amps or watts and take notes. Also, the Tri-Metric measure both AC and DC current use. The Kill-A-Watt meter can only measure AC.

Whichever method you chose, you want to measure each and every device you might use and add up the wattage values of all the items you might use all at once for a total maximum operational power usage estimate.

In my case, I use a laptop nearly all day for entertainment, email, social media and my writing. That's a maximum of 100 watts when I'm doing video editing in spite of the fact the Apple charger has 85 watts clearly printed on the side of the adapter! I also use a large LED lamp, a coffee warmer, and a fan. But on warmer days, I break out the swamp cooler. And a couple of times a day, I'll use my coffee maker, toaster oven or microwave oven. The last three are power hogs, but I only use one of them at a time. I have to because between any two of the three items would easily push my so called 3,000-watt inverter to its real-world 2,500-watt limit.

As you might have already guessed, I've reached the limit of my capacity with my choices of electrical devices, but instead of looking for some monster 5,000-watt inverter, I chose to have two separate inverters; one a pure sine wave inverter and the other a square wave or modified sine wave inverter. I did this for two reasons; 1) I actually started out with the square wave inverter and learned it wasn't enough for what I wanted to do, and 2) After learning that I needed more power capacity and my microwave oven was buzzing like crazy on the square wave inverter, I bought another pure sine inverter for my pickier equipment like the microwave and my RAID storage array for my computer. Also, at the time, since I hadn't yet learned the unavoidable need for a proper MPPT solar charge controller, I needed a battery charger. So, I bought an expensive 3,000-watt pure sine inverter/charger combo. Now, between the two inverters, I can run everything I want all at the same time (well, within reason. I still don't run the big guns at the same time like my toaster oven and the microwave. They have to take turns. Hey! I'm living off batteries here!)

One other issue I had with my inverters was the fact that my Apple laptop could not function on the more expensive pure sine inverter because of a floating ground problem. For whatever reason, the much less expensive square wave inverter handled floating ground properly so the touchpad on my laptop wouldn't make the mouse on the screen jump around erratically. I talked more about floating ground in the chapter on inverters earlier.

My point is this: If you merely double your power estimates, you take a chance at only breaking even and perhaps finding yourself without power at some point. By tripling your estimate, you give yourself a wider margin for error, but that margin will also allow you to add more appliances later or give you more time on your batteries some night while you're up late binge watching your favorite TV show episodes on your laptop.

# Battery Chargers

Right away, you need to know that the most deceptive aspect of the solar power business is battery chargers. I'm not talking about solar charge controllers. I mean the separate wall outlet powered battery chargers. There is no such thing as a "smart" charger. These chargers are dumb as rocks and absolutely worthless! They also cost more than real battery chargers because they can sell the word "smart". And this isn't me being an old fuddy-duddy about "new-fangled computers and such nonsense". No, "smart" chargers are just a flat out lie. And they've even taken that lie to a whole new level with so called "pulse" chargers. These are supposedly "smart" chargers that also pulse the charge by fluctuating the current to try to "shock" the battery into eliminating sulfation and holding a stronger charge. It's a load of crap. Don't fall for this garbage.

Batteries need a steady high voltage to charge, a steady lower voltage to float charge, and an even lower voltage to maintain an unused battery for long periods of disuse. These stages of charge are well known by the battery manufacturers and have specific names. The first stage is called bulk charge. The voltage is slowly ramped up and the current is constant. Once the battery reaches its bulk charge limit (about 14.8 volts for 12 volt systems), it changes to absorption charge mode. At this stage, the charge is maintained at a steady voltage as the current needed is reduced over time. Once the battery is fully charged, it moves to the float charge stage where the battery is maintained in a fully charged state (about 13.8 volts). These so-called smart chargers hardly ever do this right because they're programmed to be safe and idiot proof. The voltages they provide are often too low to do the job.

With a properly designed and built solar power system, your solar charger should start charging in bulk mode as soon as the sun is out and rapidly approach absorption mode by lunchtime. Float mode should come fairly soon afterwards.

Since you'll actually be using the power system and not leaving it sitting in a garage somewhere, all you need to be concerned about is making sure that your batteries never dip below 12.2 volts. The only time you need a "battery maintainer" or "smart charger" is if you park your vehicle out of the sunlight in a garage for an extended period of time. The so-called "smart" chargers are supposed to know which voltage you need at any given time, but they don't work as advertised and even if they did, they're not what you need for a solar power system. The ONLY time they might be useful is if you store your vehicle out of the sun.

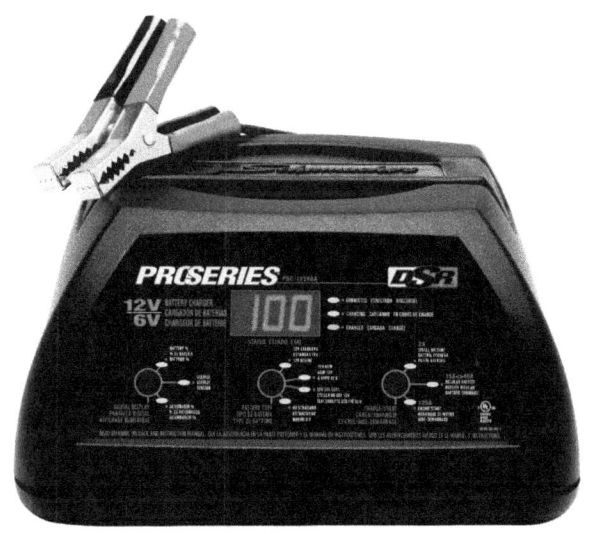

"Smart" battery charger.

With a properly installed and functioning solar power system and sufficient sunlight, you should never have to use a battery charger. But this is one of the few places you can save money if you're on a tight budget. If you're primary inverter doesn't have a built-in battery charger, you can get a constant amp battery charger for a lot less than the "smart" chargers that will give you a constant voltage and amperage to properly charge your batteries. Of course, you'll need an AC power source like a generator or a wall socket. Automotive shops and some department stores will carry a 40 or 60-amp battery charger. They'll be larger than the "smart" chargers and probably louder as well, but they'll work. You just have to watch them to make sure you don't overcharge your batteries. These chargers are even dumber than the so called "smart" chargers.

The trick is knowing how long to charge your batteries. If you have a proper battery monitor, which you should (I recommend the Tri-Metric of course), then you should be able to easily determine when your batteries are fully charged. Charging this way is cumbersome, time-consuming and (as I mentioned earlier) unnecessary if your solar power system is built properly.

**Standard automotive battery charger.**

If space is a premium and you don't have room for a large battery charger but you can afford a more expensive inverter, you could buy one with a built-in battery charger. They're usually more configurable for your battery type and the level of charge you need, and unlike the "smart" charger are actually designed fairly intelligently for what they do. They're also much more compact. However, an inverter with a built-in charger will be very heavy, so keep that in mind.

If you're considering connecting your solar battery bank to your vehicle alternator as an alternate source of charging, don't. The only way to safely and effectively charge a solar battery bank in that manner is to replace the existing alternator with a heavy-duty 150 or 200-amp inverter along with a large power isolation block separating the vehicle battery from the solar battery bank. With parts and labor, IF you can find somebody qualified to install it properly, the cost to do this could run at least $500. And once it was all installed, it would only work if the engine was running. In most states, you can't legally run a car at idle longer than 20 to 30 minutes unless it's diesel and commercially licensed and if you did, it's not good for the engine to do that anyway.

Instead, take that money and make sure you use heavy gauge wires and a proper MPPT solar charge controller. I can't stress this enough. You should NEVER have to use a generator or wall outlet powered battery charger if your solar power system is designed and build properly.

# Generators

As I've mentioned before, if your solar power system is designed and built properly, you should NEVER have to use a generator or wall outlet powered battery charger. BUT, there is one item that batteries can't run for long; an air-conditioner.

If you need to run an air-conditioner, you're going to need a generator. Technically, it is possible to run a small air-conditioner off a large battery bank, but not for long. I've done it, but the best I could get out of my six 6-volt deep cycle battery bank at full charge under direct sunlight with the charge controller actively charging the batteries was about six hours for the smallest air-conditioner you can buy (about 30 amps). The batteries dropped down to about 65% and it took two sunny days to fully recharge them. The truck never cooled down below 80° anyway because it was 90° outside, which equates to nearly 100° inside. I was also using my usual electrical devices throughout the entire time, so if push came to shove, I guess I could have just sat back in low light and read a book and maybe captured another couple of hours from the air-conditioner if I had to.

A slightly better option is to buy a portable swamp cooler. Of course, they work better in dry, arid conditions and don't cool as well as an air-conditioner, but they use less power doing it. Mine is huge and uses about 23 amps. Compare that to a large AC powered fan at top speed which uses only 4 amps.

If you must buy a generator, there is only one choice; Honda. Period. That's it. Yes, they are the most expensive generator you can buy, but there's a reason for that. They are also the best quality, the smallest for their wattage ratings, the most reliable, the most versatile in terms of expandability and options, the easiest to get repaired and serviced, and most importantly, they're very quiet.

**Honda EU2000I Super Quiet Generator.**

Some might argue that Generac makes a quieter generator for a lower price. And they may be right, but they lose points for build quality, reliability, versatility, and getting them repaired is not near as easy.

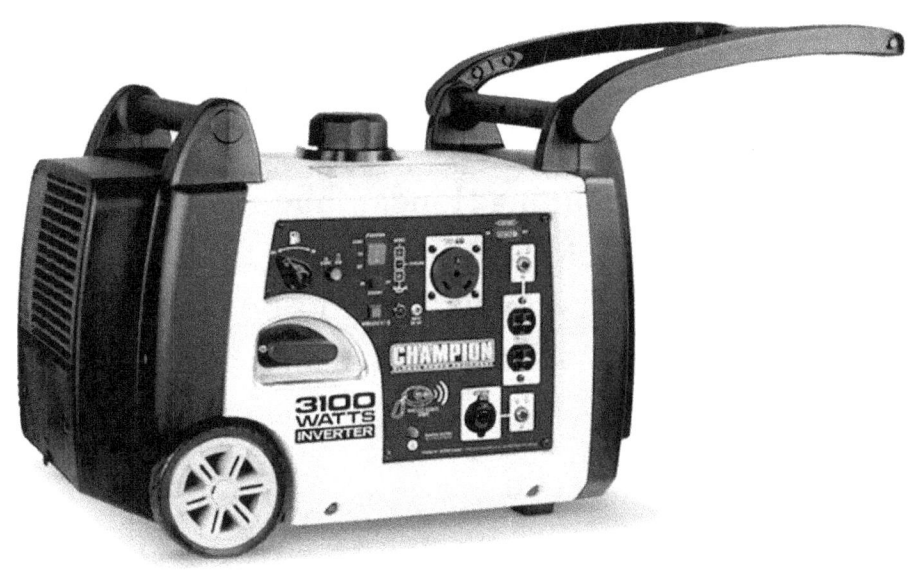

**Champion 3,100 watt generator.**

A distant second option is a brand called Champion. I own a Champion generator. I couldn't afford a Honda at the time and I felt I needed a more powerful generator for my money because I hadn't properly built my solar power system in the first place. Now that I have a proper MPPT charger controller and the right heavy gauge wires as short as possible, I have no need for the generator at all. It's just a big yellow brick in the way, about twice the size of a decent Honda model. The only upside of my purchase is that for the same amount of money, I got 3,100 watts for what would only have gotten me 2,000 on a Honda. But since you only need a 2,000 watt to run a small air-conditioner anyway, that's ideal.

My Champion is also twice the size of the Honda, twice as heavy, and slightly louder. Not nearly as loud as a diesel generator might be or the monster propane-powered model I initially bought at Costco that was so loud, I ran it for literally three seconds, shut it off, and promptly returned it.

The best way I can describe the difference in noise between my Champion and a Honda is in feet distance before you can't hear it anymore. A properly running Honda while still under heavy use will be virtually undetectable about thirty feet away. My Champion is about 50 feet away. For comparison, most other generators can be heard across an entire Walmart parking lot!

# Capacitors

Oooh! This is one seriously, hotly debated and contested subject. Some people have posted videos on YouTube showing the connection of these high-volume capacitors in their power systems to help provide a more stable and reliable power source to the inverter. They swear that it improves power flow and allows them to use power-hungry devices like freezers or table saws without severely taxing the power system.

**80 farad automotive sound system capacitor.**

There are also people who will hotly deny any benefit at all to connecting a large capacitor in parallel to your battery/inverter connection. They'll scream in all-caps on the Internet that it's utter hogwash. And I laugh my tail off as I drink my tankard of hot coffee that I just brewed on my Keurig K-cup instant coffee maker using my 3,000-watt inverter and an 80 farad capacitor in parallel with the six 6-volt golf cart batteries in my truck.

My battery bank is wired for 12 volts; three parallel connected two-battery serial pairs. They're all 6-volt deep cycle golf cart batteries, so I have to connect two of them in series to get 12 volts and then all six of them in parallel twins for a collective total of about 600 amp/hours at 12 volts. Yet, with all of that power at full charge and the sun beaming down directly on me, I could never use my Keurig coffee maker. It's a beast. It's basically a dead short in a bucket of water to heat it to boiling in roughly 145 seconds. It draws about 1,600 continuous watts and it draws it so fast that the batteries simply cannot deliver it fast enough, even with the huge 1/0 gauge wires I use from the batteries to the inverter. But after installing a massive 80 farad capacity that is ordinarily used for automotive sound systems to prevent the heavy base from the car stereo from dimming the headlights every time there's a sick beat, I was able to easily run my coffee maker.

The capacitor works like a quick charging, fast discharging battery. It provides another source of 12 volts in parallel with the battery bank. The difference is, the capacitor provides that 12 volts as fast as the coffee maker demands it. 1,600 continuous watts for a couple of minutes is a lot of power! It's so much in fact, that after brewing one 20-ounce tankard of coffee, my entire battery bank is depleted by 1%. You might do the math and calculate that 1,600 watts should be possible, but I'm telling you from real-world daily coffee drinking, my coffee maker doesn't work without the capacitor but does with it. That's a perfect example of theory verses real-word experience. Take it or leave it - I'm going to enjoy my coffee with a smile on my face.

It is possible that if I upgraded my 1/0 wires from the battery bank to the inverter to 2/0 or even 3/0, I might be able to get enough power to the coffee maker without the capacitor. But the 1/0 wires already require such massive loop terminals on the ends, I would be hard-pressed to get anything larger to fit on the battery and inverter terminals. Also, in hindsight, the capacitor terminals can only fit up to 1/0 wire as it is. So I had to decide on spending a lot of money on heavier wire and foregoing the capacitor, or staying with what I already had installed and install the capacitor. Besides, I really wanted to experiment with the capacitor to be able to prove or disprove their usefulness for this book. So...you're welcome.

# Battery Resuscitation Tricks

Right off the bat, forget about adding aspirin to flooded acid batteries. That's an old myth. You've really only got two choices if you even what to try to save a near dead battery; Epsom salt and distilled water or a baking soda bath and new acid. And forget about so called "smart" or "pulse" battery chargers that supposedly "shock" the sulfation off the battery cells. That's a load of crap. They don't work.

Sulfation is what kills batteries. Sulfation is a crystalline substance made of combined lead and sulfuric acid to form a light purple powdery substance that builds up on the plates in the cells. Once a battery becomes overly sulfated, it quickly starts to lose charging capacity. Eventually, the sulfation will build up enough to short out the battery cells and that's the end of the road. You can't resuscitate a shorted out battery. They're truly dead. But even a heavily sulfated battery can be resuscitated (at least for a while) to keep you going until you can get new batteries.

Quick note: Never combine new batteries with old ones in a battery bank, especially if the old ones are nearly dead. The new batteries will be constantly trying to charge the older ones and your battery bank will function at a fraction of its potential capacity. In fact, it only takes one dead cell in one battery to cause the whole battery bank to work at half efficiency!

There are two schools of thought on resuscitating flooded acid batteries; stay true or go rogue. The "stay true" crowd tries to remove the sulfation by draining the acid from the batteries and then washing the cells with an ample baking soda and distilled water mixture several times until the wash comes out as clear as possible. Then once the cells are as clean as possible and allowed to dry out a bit, they fill the batteries with all new acid. This approach is messy, difficult, and legally complicated. It's messy because you're dealing directly with battery acid coming out and going back in and well as a slurry of baking soda, water and lead deposits actively foaming out of the cells. You need a large catch basin, a five gallon bucket, rubber gloves, eye protection and clothes you don't care to keep. The batteries are also extremely heavy and you have to lift and overturn them several times during the process.

The "go rogue" crowd dumps the acid out and replaces it with Epsom salt and distilled water without a wash out to clear the sulfation deposits. They avoid some of the battery lifting, but they still have to deal with disposing the old acid.

Disposing old acid is tricky business. Some states have more stringent regulations than others. One safe way to do it is to put it all in a five-gallon bucket and slowly stir in enough baking soda until it becomes an inert mud. That neutralizes the acid well enough, but you still have a hazardous material issue to deal with.

Either way you might choose, once you wire the batteries back together into a full 12-volt battery bank, you follow up with a good four hour desulfation charge cycle at 15.5 volts to loosen up the remaining deposits on the plates and give the poor batteries at least a fighting chance. The batteries will outgas a bit while the heat evaporates and "boils off" some of the water, so keep an eye on water levels to make sure the plates don't get exposed to the air.

One important thing to keep in mind about changing the chemistry of a battery with Epsom salt instead of acid is that the gas that escapes from the battery during heavy charge will no longer be the hydrogen you would expect from a standard acid flooded battery. Instead, the salt causes chlorine gas. Sound unpleasant? Yeah. It can cause respiratory problems and at the very least sting your eyes.

And here's the punchline. Regardless of which technique you choose, the best you might get out of these near dead batteries after all that heavy lifting and working with dangerous chemicals is maybe six to eight months tops before they die completely. Considering the cost of equipment, materials, and unavoidably damaged clothing, you're better off just buying new batteries or at least professionally refurbished ones.

The absolute best way to deal with the problem of sulfation is to not let it happen in the first place. And doing that is a lot simpler than you might expect. All you have to do is three things; 1) Make sure the distilled water levels in the cells never drop below the surface of the plates, 2) Design and build a proper solar charging system with an MPPT charge controller that routinely goes through the proper charging modes each day at the correct voltages, and 3) Run a desulfation charge mode at 15.5 volts for at least 3 to 4 hours once every month. That's it. A good MPPT charge controller will have the charging modes and voltages programmed by default along with a regularly scheduled desulfation charge every 28 or 30 days, so all you really have to do is keep an eye on the water levels. Easy.

Now, if you have sealed AGM (no-maintenance) batteries or lithium batteries, all of this unnecessary. You only need to know this for the cheaper flooded lead acid batteries.

I had a bank of six 6 volt batteries that were never fully charged because I had a worthless PWM solar charge controller. The sulfation gradually built up on the cell plates until finally, I couldn't get through a day without the entire battery bank giving out on me. At the time, I didn't have the money to buy a whole new set of batteries. So, I did my research, asked a friend whose been down this road himself and finally decided to "stay true" to acid for two reasons; 1) I liked the idea of actually washing out the sulfation deposits in spite of all the heavy lifting, and 2) I preferred the idea of gassing hydrogen instead of chlorine in the enclosed cab of my truck. I also already had a large catch basin normally used to put under hot water heaters that I had purchased to put under a leaking portable chemical toilet (that's another story). And I had the rubber gloves I used for when I used to dump the black water on my old RV which I no longer had (yet another story). I didn't use any eye protection beyond my usual glasses to see what the heck I'm doing. (I don't recommend that by the way.) After I was done with this little experiment, I literally had to throw away ALL of my clothes because of the acid splashes. I'm lucky I didn't splash acid into my eyes.

After all the work and heavy lifting that took three whole days to get done, not to mention the cost of new acid that I could only get at a larger warehouse auto parts shop across town (at eight miles to the gallon on the truck), I had spent 72 hours, 2 smashed fingers, one set of jeans and a cheap t-shirt, $20 for gas, and about $30 for new acid. And in the end, the batteries worked only about 75% of full capacity for another six months before completely giving up the ghost. Luckily, that was enough time for me to save up for six new refurbished batteries.

My buddy suggested I "go rogue" instead with the Epsom salt solution. I was already against the idea for the lack of sulfation wash out and resulting chlorine gas when I caught a local acquaintance of mine cursing and throwing his batteries into a garbage bin nearby. When I asked him what was troubling him, he explained that he had tried to save his batteries with Epsom salt and they only lasted about eight months. When I confronted him about properly disposing of the batteries at an auto parts shop instead of tossing them into the garbage, he argued that they were just heavy boxes of salt now, not acid. I thought to argue about the lead issue, but he wasn't in a mood for debate. He's also four inches taller than me, about 50 pounds heavier, and ten years younger. So, I waited until he left and retrieved the batteries out of the garbage bin to see if I could possibly revive them or at least dispose of them properly. I took them to my favorite auto parts store where they tested them and they all came back dead as a doornail. I at least felt justified for keeping all that lead out of the landfill. Besides, can you imagine the noise that would make when the garbage bin got emptied?

Bottom line, as is always the case with solar power systems, do it right from the beginning and this will all be a moot issue. Of course, the trick to doing it right the first time is getting useful advice in the first place. That's why I wrote this book.

# Basic Electronics

You should notice that this section will be brief and lacking any serious detail. That's because if I have to teach you electronics, this is going be a much bigger book. This section is just supposed to be an introduction to some basic concepts, just in case you know nothing about electronics. The information provided here is not so much to teach you about electronics as it is to give you some new vocabulary in context so you might recognize it while reading other sections in the book.

Also, I cannot be held responsible for any shocks, fires, injuries or firework shows that may result from misusing any information in this book.

Electronics can be summarized with just a couple of simple formulas:

**Voltage = Current x Resistance**

**Wattage = Current x Voltage**

Current is measured in amps. Resistance is measured in ohms. And Voltage is measured in volts. Yeah, okay, duh. So what?

Well honestly, the only thing you might really need to know is the wattage formula. We used it in the section on Fuses in the chapter on Wiring. It works with both AC and DC voltages. Just be aware that the power a device like an inverter produces is not the same as the power it uses in order to do it. The same is true of microwave ovens. A 1,000-watt microwave uses much more than 1,000 watts of power to run it. The 1,000-watt rating is the amount of power being delivered to the food inside.

Amps and volts work differently in series and parallel circuits. Amps increase in parallel circuits, but volts increase in series circuits. Crazy, huh? This kind of makes sense though. They're basically on different sides of the equation. In fact, the job of a proper MPPT solar charger is to trade volts for current and vice-versa when needed.

This all becomes very important when dealing with wire thickness or gauge. Think of a wire as a water hose, the current as the water (that's easy enough), and the voltage as the water pressure in the hose. A thin wire cannot handle as much as a thick wire when it comes to current and voltage. The easy answer as to why this is true is that there are simply fewer atoms in the wire to transfer the charge.

If you try to run too much current through a thin wire, just like too much water through a hose will make the hose burst, too much current through a wire will cause the wire to overheat and possibly even catch fire.

While you can run a lot of voltage through a wire, voltage without current is not a lot of power. Remember the wattage formula; Watts = Amps x Volts. A perfect example of this is a Taser gun. They send hundreds of volts down tiny little wires to shock the target person. But with so little current (amps), that shock won't kill them.

All of this becomes important when installing solar power systems because if you use too small of wire (10 gauge or less), it can't handle enough current in amps to be of any use. And if you try to force it to, it can heat up the wires and cause all kinds of damage.

The other element of the Voltage equation is Resistance. Resistance is in everything. Even highly conductive wire has some resistance. And the longer the wire, the more resistance. Resistance consumes power. And whenever that happens, heat occurs. We already discussed resistance from thin wires, but there are a lot of other things that can cause resistance in a power system. In fact, the amount of power lost to heat along overhead power lines between the power station and your house can be as much as 15% depending on how far and how old the cables are.

Loose connections and connections between different types of metals will result in increased resistance as well. Connecting wires with crimps instead of soldering them with a soldering iron can create loose connections between two different kinds of metal. Oil and dirt between connections, particularly on battery terminals can create resistance. Even if you made sure to keep your connections clean when you first installed everything, if they get exposed to the weather or hydrogen outgassed from lead acid batteries, they can develop corrosion which results in resistance.

In the end, resistance reduces power capacity. So always make sure your connections are clean and preferably soldered and not just crimped or twisted together. This is also important to avoid accidental shorts from loose wires.

There's always room for more egghead talk about electronics, but the bottom line is, if you don't know enough to feel comfortable working with all the elements of a solar power system, by all means find someone to help you. You can be seriously injured by the kind of power that comes out of a single battery or the leads coming from a solar panel array on even a cloudy day. No joke.

# Example System Diagrams

## Simple/Minimal System

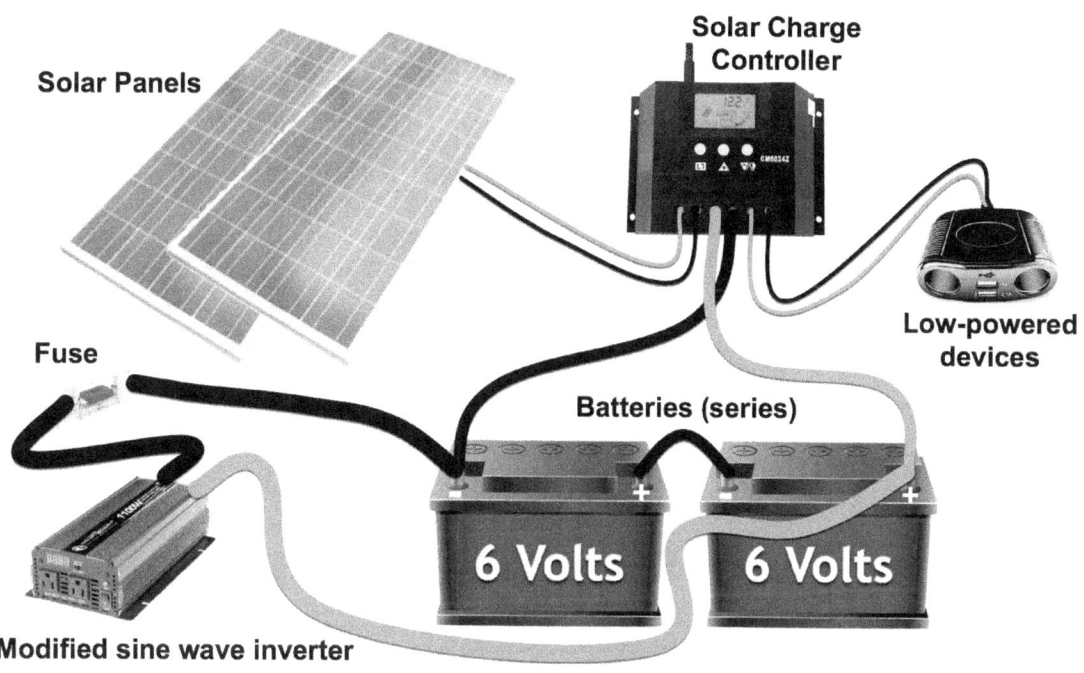

The absolute minimum system would include one or two 100 watt solar panels, so you can get away with a cheap PWM solar charge controller because the panels are the same size, type and less than 500 watts total. You'll want at least an 800-watt modified sine inverter and two 6-volt deep cycle batteries wired in series for a 12-volt system. The PWM controller will also likely have a "load" terminal where you can connect low-powered DC devices like cellphone chargers or fans or LED lights. Notice the size of the wires for each connection; 10-gauge standard from the solar panels and from the controller load terminals, 4 or 6 gauge from the controller to the batteries and 1/0 gauge from the batteries through the fuse to the inverter.

This size of power system should provide enough power to run a laptop, small TV, 12 VDC electric blanket or fan, LED light and a cellphone charger. That's about it. But that's pretty good.

You should also notice that the inverter is connected to the battery bank at opposite terminals of the two batteries.

# Day-to-Day Functional System

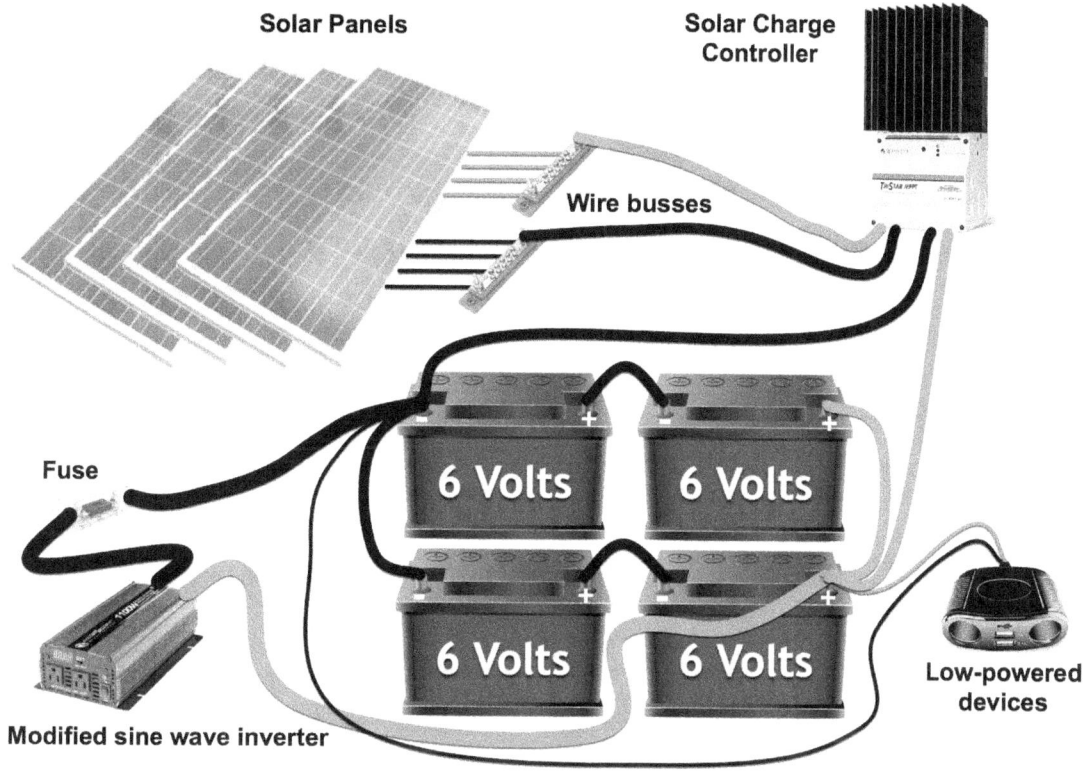

At this level of functionality, you have to spend a bit of money. For the kind of power you're going to need on a daily basis, you absolutely must spend the extra money for a proper MPPT solar charge controller. Along with that comes more panels, more batteries, and a bigger inverter.

For the kind of power to run all the items mentioned above with the minimal system in addition to perhaps a coffee maker, rice cooker, crockpot, or toaster (not at the same time of course), you'll need at least 500 watts of solar panel power, four 6 volt batteries wired in series and parallel to make a 12-volt battery bank, and a 1,500 or 2,000 watt inverter. Since you're still not running anything particularly sensitive like a microwave oven, you can save some money by sticking with a modified sine wave inverter instead of a much more expensive pure sine wave inverter.

Most high-end MPPT solar charge controllers don't provide "load" terminals for low-powered 12 volt devices, so you'll have to connect to the battery bank directly for your DC devices.

Here again, the inverter is connected to the battery bank at opposite terminals across the entire battery bank.

# Robust "Road Warrior" System

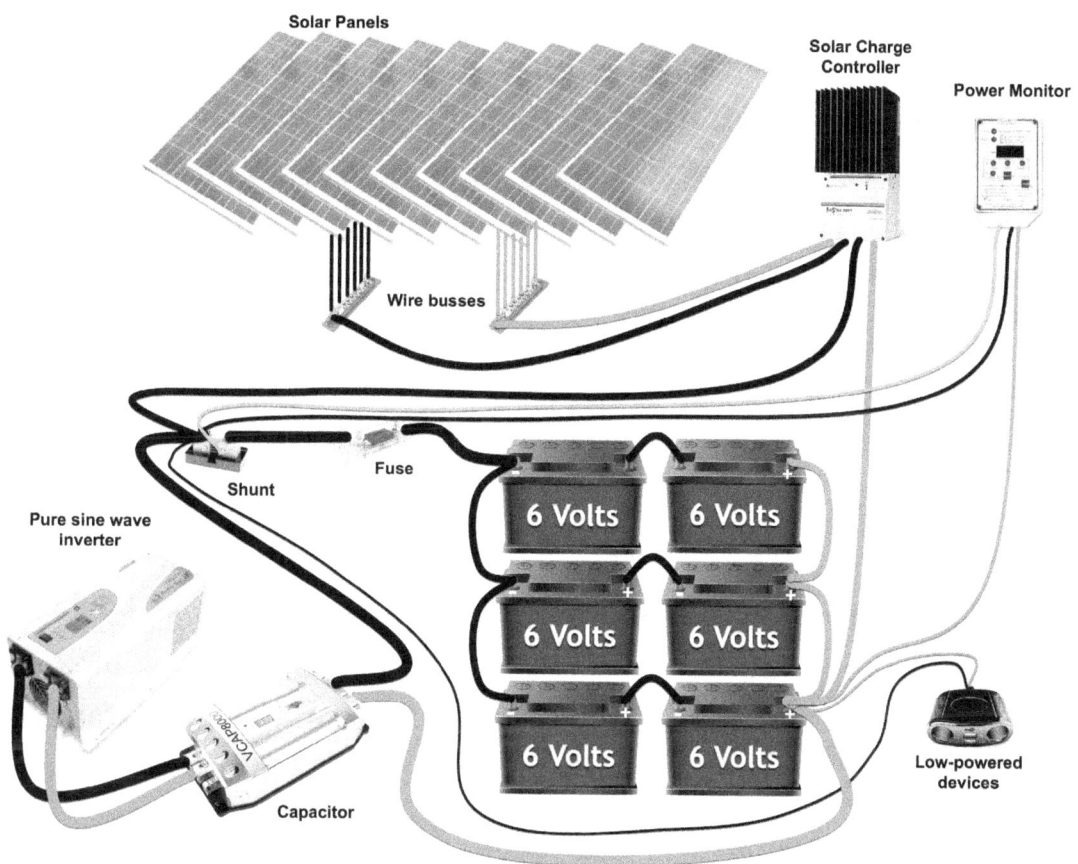

This level of power is robust enough to allow any mobile nomad to live almost as comfortably as if in their own house connected to power lines. The driving force in this scenario is basically the ability to run a microwave oven or instant coffee maker at full power without any worries - or the equivalent amount of power over time in order to last through several cloudy or rainy days in places like Oregon during the rainy season.

You'll need at least 1,000 watts of solar panel power, but the truth is if there's space on your roof for more, cover it with solar cells. There's no such thing as too much solar.

You'll also need a larger battery bank; at least six 6 volt batteries wired in series and parallel for a 12-volt battery bank. That much power requires a proper MPPT solar charge controller as well. And since you're connecting several solar panels in parallel to form a high current array, you must connect the solar panels to a wire buss connected to larger 4 or 6-gauge wire to connect to the solar charger. Similarly, the kind of power you'll be pulling from the batteries in order to run a microwave at full power will require some seriously heavy wire from the batteries to the inverter(s).

Since you're running a microwave, you absolutely must use a pure sine wave inverter as well. These can be very expensive when compared to the price of a modified sine inverter, especially since you'll need a good 3,000-watt inverter. And if you're really interested in pushing the limits of the system with an instant coffee maker, you'll need a capacitor between the battery bank and the inverter.

Finally, with a system this powerful, you're going to need to know just how much power you actually have. A simple light bar meter or analog volt gauge won't cut it. You'll need a proper battery power monitor.

And, as always, the inverter is connected to the battery bank at opposite terminals across the entire battery bank.

With a system like this, you'll be able to do pretty much whatever you want as long as you don't try to run too many high current devices at once like a microwave and a toaster or coffee maker at the same time. You just have to use each in order. Sure, it can slow up making breakfast, but that breakfast includes toast and coffee. It's not just a bowl of cold cereal with milk.

With the addition of another 100 watts of solar cells and a second inverter, this is the very configuration I live with every day in my truck and I have no complaints whatsoever.

# Sources for Materials and Equipment

**Best batteries:**

    CROWN BATTERIES: crownbattery.com

    TROJAN BATTERIES: trojanbattery.com

**Best Battery Monitors:**

    TRI-METRIC: bogartengineering.com

**Best solar charge controllers:**

    TRISTAR: morningstarcorp.com

**Best solar equipment dealers:**

    BACKWOODS SOLAR: backwoodssolar.com

    DONROWE.COM: donrowe.com

**Best Internet Blog:**

    HANDYBOB'S BLOG: handybobsolar.wordpress.com

www.ingramcontent.com/pod-product-compliance
Lightning Source LLC
Chambersburg PA
CBHW082331220526
45470CB00008B/2474